DNA and RNA: Understanding Genetics

DNA and RNA: Understanding Genetics

Samantha Matthews

MURPHY & MOORE
www.murphy-moorepublishing.com

DNA and RNA: Understanding Genetics
Samantha Matthews
ISBN: 978-1-63987-168-1 (Hardback)

MURPHY & MOORE
Published by Murphy & Moore Publishing,
1 Rockefeller Plaza,
New York City, NY 10020, USA

Cataloging-in-Publication Data

DNA and RNA : understanding genetics / Samantha Matthews.
p. cm.
Includes bibliographical references and index.
ISBN 978-1-63987-168-1
1. DNA. 2. RNA. 3. Genes. 4. Nucleic acids. 5. Genetics. I. Matthews, Samantha.
QP624 .D53 2022
572.86--dc23

For more information regarding Murphy & Moore Publishing and its products, please visit the publisher's website www.murphy-moorepublishing.com

Table of Contents

Preface

This book has been written, keeping in view that students want more practical information. Thus, my aim has been to make it as comprehensive as possible for the readers. I would like to extend my thanks to my family and co-workers for their knowledge, support and encouragement all along.

Genetics is the study of genes, heredity and genetic variation in living organisms. RNA and DNA are nucleic acids and form two of the four major macromolecules essential for all known life forms. DNA carries genetic information that is important for the functioning, development, reproduction and growth of organisms. It is formed by two chains that are coiled around each other forming a double helix whereas RNA is a single stranded molecule. RNA plays an important role in coding, decoding, regulation and expression of genes. It is used to convey genetic information and transmit cellular signals. It carries information from the DNA to ribosome where protein is synthesized in the cell. This book is compiled in such a manner, that it will provide an in-depth knowledge about the theory and practice of this discipline. It also elucidates the concepts and innovative models around prospective developments with respect to the field of genetics. Coherent flow of topics, student-friendly language and extensive use of examples make this book an invaluable source of knowledge.

A brief description of the chapters is provided below for further understanding:

Chapter - Introduction to DNA

The molecule which is composed of two chains which coil around each other to form a double helix that carry genetic instructions is known as Deoxyribonucleic acid or DNA. It is responsible for the functioning, development, growth and reproduction of all known organisms and many viruses. This chapter has been carefully written to provide an easy understanding of DNA.

Chapter - Understanding RNA

The polymeric molecules which have vital biological roles in coding, decoding, regulation and expression of genes is known as RNA. The types of RNA include m-RNA, t-RNA, r-RNA and tmRNA. The topics elaborated in this chapter will help in gaining a better perspective about the functions, properties, and types of RNA.

Chapter - DNA Replication

DNA replication refers to the process of producing a copy of DNA itself during cell division. DNA-binding proteins are proteins having DNA-binding domains and general affinity for single- or double-stranded DNA. This chapter will briefly introduce all the significant aspects of DNA replication.

Chapter - DNA Damage and Repair

The alteration in the chemical structure of DNA which causes changes in the structure of genetic material is referred to as DNA damage. DNA repair refers to the processes by which a cell identifies and cures damage to the DNA molecules which encode its genome. The topics elaborated in this chapter will help in gaining a better perspective about DNA damage and repair.

Chapter - DNA Profiling and Sequencing

The process of determining the DNA characteristics of an individual is referred to as DNA profiling. DNA sequencing refers to the process of determining the nucleic acid sequencing in DNA. All the diverse aspects of DNA profiling and sequencing including their applications have been carefully analyzed in this chapter.

Chapter - Applications of DNA and RNA

DNA and RNA find their applications in various fields such as genetic engineering, agriculture, DNA origami, DNA nanotechnology, and CRISPR and associated systems, gene expression profiling, identification of allele-specific expression, gene fusions, etc. All these diverse applications of DNA and RNA have been thoroughly discussed in this chapter.

Samantha Matthews

1

Introduction to DNA

The molecule which is composed of two chains which coil around each other to form a double helix that carry genetic instructions is known as Deoxyribonucleic acid or DNA. It is responsible for the functioning, development, growth and reproduction of all known organisms and many viruses. This chapter has been carefully written to provide an easy understanding of DNA.

Deoxyribonucleic acid (DNA) is a nucleic acid that contains the genetic instructions used in the development and functioning of all known living organisms. The main role of DNA molecules is the long-term storage of information. DNA is often compared to a set of blueprints, since it contains the instructions needed to construct other components of cells, such as proteins and RNA molecules. The DNA segments that carry this genetic information are called genes, but other DNA sequences have structural purposes, or are involved in regulating the use of this genetic information.

Chemically, DNA is a long polymer of simple units called nucleotides, with a backbone made of sugars (deoxyribose) and phosphate groups joined by ester bonds. Attached to each sugar is one of four types of molecules called bases. It is the sequence of these four bases along the backbone that encodes information. This information is read using the genetic code, which specifies the sequence of the amino acids within proteins. The code is read by copying stretches of DNA into the related nucleic acid RNA, in a process called transcription. Most of these RNA molecules are used to synthesize proteins, but others are used directly in structures such as ribosomes and spliceosomes. RNA also serves as a a genetic blueprint for certain viruses.

Within cells, DNA is organized into structures called chromosomes. These chromosomes are duplicated before cells divide, in a process called DNA replication. Eukaryotic organisms such as animals, plants, and fungi store their DNA inside the cell nucleus, while in prokaryotes such as bacteria, which lack a cell nucleus, it is found in the cell's cytoplasm. Within the chromosomes, chromatin proteins such as histones compact and organize DNA, which helps control its interactions with other proteins and thereby control which genes are transcribed. Some eukaryotic cell organelles, mitochondria and chloroplasts, also contain DNA, giving rise to the endosymbionic theory that these organelles may have arisen from prokaryotes in a symbionic relationship.

The identification of DNA, combined with human creativity, has been of tremendous importance not only for understanding life but for practical applications in medicine, agriculture, and other areas. Technologies have been developed using recombinant DNA to mass produce medically important proteins, such as insulin, and have found application in agriculture to make plants with desirable qualities. Through understanding the alleles that one is carrying for particular genes, one can gain an understanding of the probability that one's offspring may inherent certain genetic disorders, or one's own predisposition for a particular disease. DNA technology is used in forensics, anthropology, and many other areas as well.

DNA and the biological processes centered on its activities (translation, transcription, replication, genetic recombination, and so forth) are amazing in their complexity and coordination. The presence of DNA also reflects on the unity of life, since organisms share nucleic acids as genetic blueprints and share a nearly universal genetic code. On the other hand, the discovery of DNA has at times led to an overemphasis on DNA to the point of believing that life can be totally explained by physico-chemical processes alone.

Structure and Functions of DNA

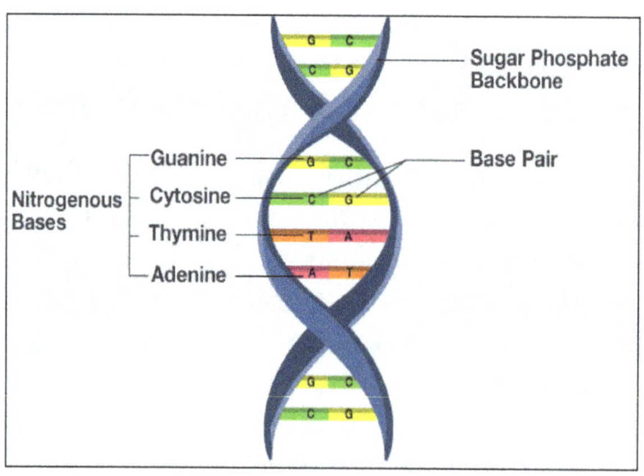

DNA Diagram showing the double-helical DNA structure.

The DNA structure can be thought of like a twisted ladder. This structure is described as a double-helix, as illustrated in the figure above. It is a nucleic acid, and all nucleic acids are made up of nucleotides.

The DNA molecule is composed of materials called nucleotides, and each nucleotide is composed of three different components such as sugar, phosphate groups, and nitrogen bases.

The basic building blocks of DNA are nucleotides, which are composed of a carbon-sugar group, a phosphate group, and a nitrogen base. The sugar and phosphate groups link the nucleotides together to form each strand of DNA. Adenine (A), Thymine (T), Guanine (G) and Cytosine (C) are four types of nitrogen bases.

These 4 Nitrogenous bases pair together in the following way: A with T, and C with G. These base pairs are essential for the DNA's double helix structure, which resembles a twisted ladder.

The order of the nitrogenous bases determines the genetic code or the DNA's instructions.

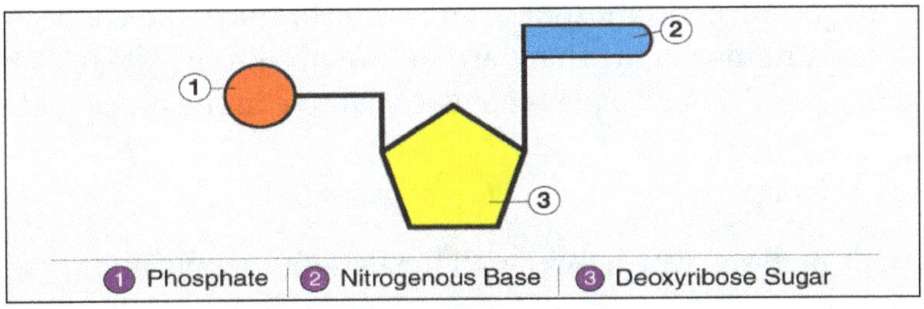

Components of DNA Structure.

Among the three components of DNA structure, sugar is the one which forms the backbone of the DNA molecule. It is also called deoxyribose. The nitrogenous bases of the opposite strands form hydrogen bonds, forming a ladder-like structure.

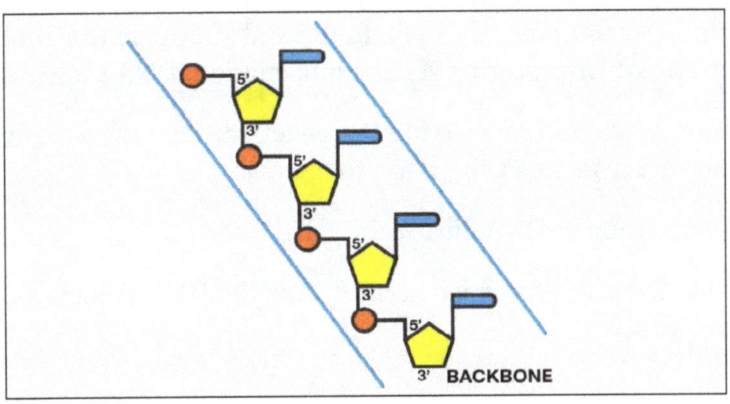

DNA Structure Backbone.

The DNA molecule consists of 4 nitrogen bases, namely adenine (A), thymine (T), cytosine (C) and Guanine (G) which ultimately forms the structure of a nucleotide. The A and G from a pair of purines and the C and T from pyrimidines. Both purines and pyrimidines differ as they have the carbon rings either single or double.

The two strands of DNA run in opposite directions. These strands are held together by the hydrogen bond that is present between the two. The strands are helically twisted, where each strand forms a right-handed coil and ten nucleotides make up a single turn.

The pitch of each helix is 3.4 nm. Hence, the distance between two consecutive base pairs (i.e., hydrogen-bonded bases of the opposite strands) is 0.34 nm.

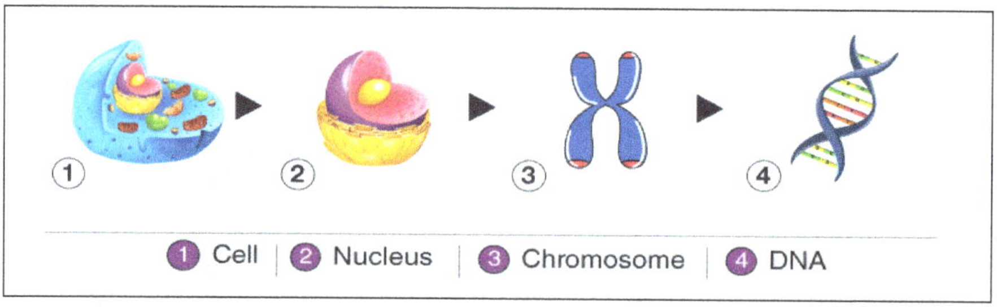

The DNA coils up, forming chromosomes, and each chromosome has a single molecule of DNA in it. Overall, human beings have around twenty-three pairs of chromosomes in the cells of the nucleus. DNA also plays an essential role in the process of cell division.

Functions of DNA

DNA is the genetic material which carries all the hereditary information which are coded in the arrangement of its nitrogen bases. Genes are the small segments of DNA, consisting mostly of 250 – 2 million base pairs, depending on the gene. A gene code for a polypeptide molecule, where three nitrogenous bases stand for one amino acid.

Thus, the sequence of a gene can be used to make a polypeptide, which then forms a protein. As every organism contains many genes in their DNA, different types of proteins can be formed. Proteins are the main functional and structural molecules in most of the organisms. Apart from storing genetic information, DNA is also involved in:

- Replication process: Transferring the genetic information from one cell to its daughters and from one generation to the next.
- Equal distribution of DNA during the cell division.
- Mutations: The changes which occur during the DNA sequences.
- Transcription.
- Cellular Metabolism.
- DNA Fingerprinting.
- Gene Therapy.

A DNA Molecule Consists of Two Complementary Chains of Nucleotides

A DNA molecule consists of two long polynucleotide chains composed of four types of nucleotide subunits. Each of these chains is known as a DNA chain, or a DNA strand.

Hydrogen bonds between the base portions of the nucleotides hold the two chains together. nucleotides are composed of a five-carbon sugar to which are attached one or more phosphate groups and a nitrogen-containing base. In the case of the nucleotides in DNA, the sugar is deoxyribose attached to a single phosphate group (hence the name deoxyribonucleic acid), and the base may be either adenine (A), cytosine (C), guanine (G), or thymine (T). The nucleotides are covalently linked together in a chain through the sugars and phosphates, which thus form a "backbone" of alternating sugar-phosphate-sugar-phosphate. Because only the base differs in each of the four types of subunits, each polynucleotide chain in DNA is analogous to a necklace (the backbone) strung with four types of beads (the four bases A, C, G, and T). These same symbols (A, C, G, and T) are also commonly used to denote the four different nucleotides—that is, the bases with their attached sugar and phosphate groups.

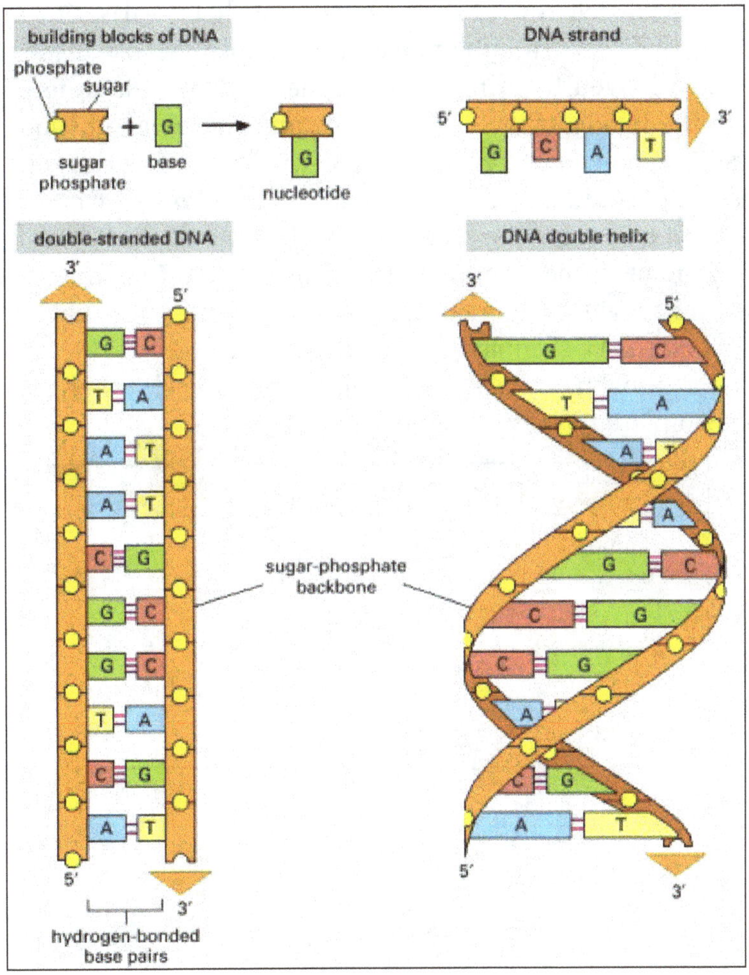

DNA and its building blocks.

DNA is made of four types of nucleotides, which are linked covalently into a polynucleotide chain (a DNA strand) with a sugar-phosphate backbone from which the bases (A, C, G, and T) extend. A DNA molecule is composed of two DNA strands held together by hydrogen bonds between the paired bases. The arrowheads at the ends of the DNA strands indicate the polarities of the two strands, which run antiparallel to each other in the DNA molecule. In the diagram at the bottom left of the figure, the DNA molecule is shown straightened out; in reality, it is twisted into a double helix, as shown on the right.

The way in which the nucleotide subunits are lined together gives a DNA strand a chemical polarity. If we think of each sugar as a block with a protruding knob (the 5′ phosphate) on one side and a hole (the 3′ hydroxyl) on the other, each completed chain, formed by interlocking knobs with holes, will have all of its subunits lined up in the same orientation. Moreover, the two ends of the chain will be easily distinguishable, as one has a hole (the 3′ hydroxyl) and the other a knob (the 5′ phosphate) at its terminus. This polarity in a DNA chain is indicated by referring to one end as the 3′ end and the other as the 5′ end.

The three-dimensional structure of DNA—the double helix—arises from the chemical and structural features of its two polynucleotide chains. Because these two chains are held together by hydrogen bonding between the bases on the different strands, all the bases are on the inside of the double helix, and the sugar-phosphate backbones are on the outside. In each case, a bulkier two-ring base (a purine;) is paired with a single-ring base (a pyrimidine); A always pairs with T, and G with C. This complementary base-pairing enables the base pairs to be packed in the energetically most favorable arrangement in the interior of the double helix. In this arrangement, each base pair is of similar width, thus holding the sugar-phosphate backbones an equal distance apart along the DNA molecule. To maximize the efficiency of base-pair packing, the two sugar-phosphate backbones wind around each other to form a double helix, with one complete turn every ten base pairs.

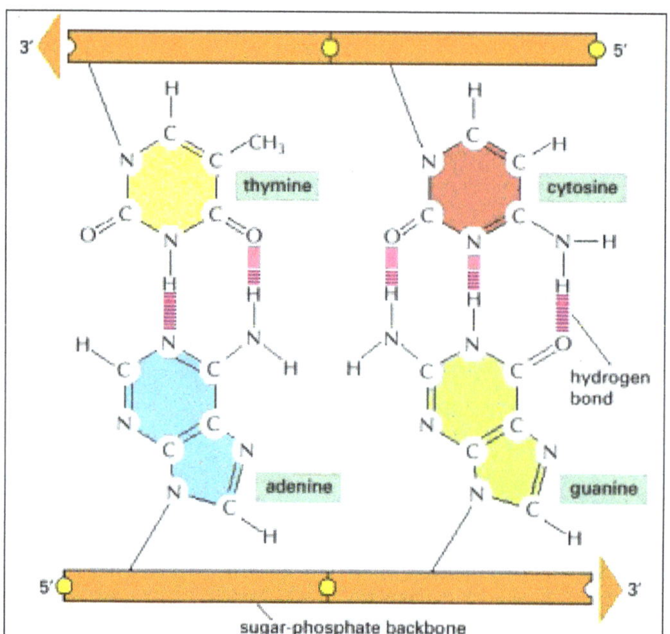

Complementary base pairs in the DNA double helix.

The shapes and chemical structure of the bases allow hydrogen bonds to form efficiently only between A and T and between G and C, where atoms that are able to form hydrogen bonds can be brought close together without distorting the double helix. As indicated, two hydrogen bonds form between A and T, while three form between G and C. The bases can pair in this way only if the two polynucleotide chains that contain them are antiparallel to each other.

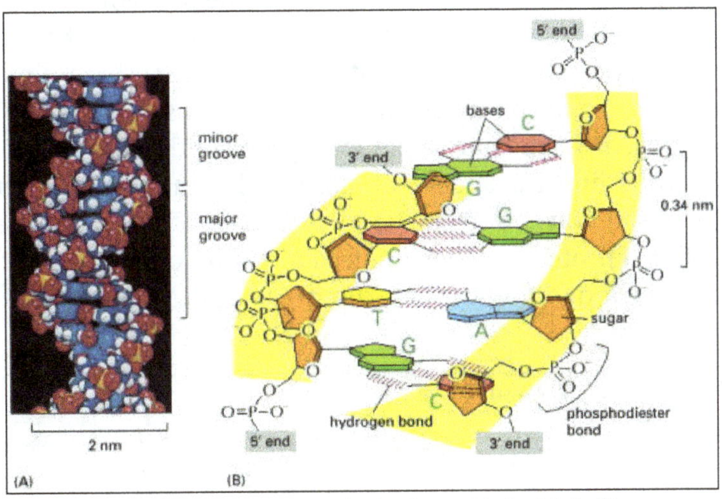

Complementary base pairs in the DNA double helix.

The shapes and chemical structure of the bases allow hydrogen bonds to form efficiently only between A and T and between G and C, where atoms that are able to form hydrogen bonds can be brought close together without distorting the double helix. As indicated, two hydrogen bonds form between A and T, while three form between G and C. The bases can pair in this way only if the two polynucleotide chains that contain them are antiparallel to each other.

The members of each base pair can fit together within the double helix only if the two strands of the helix are antiparallel—that is, only if the polarity of one strand is oriented opposite to that of the other strand. A consequence of these base-pairing requirements is that each strand of a DNA molecule contains a sequence of nucleotides that is exactly complementary to the nucleotide sequence of its partner strand.

Genes carry biological information that must be copied accurately for transmission to the next generation each time a cell divides to form two daughter cells. Two central biological questions arise from these requirements: how can the information for specifying an organism be carried in chemical form, and how is it accurately copied? The discovery of the structure of the DNA double helix was a landmark in twentieth-century biology because it immediately suggested answers to both questions, thereby resolving at the molecular level the problem of heredity.

DNA encodes information through the order, or sequence, of the nucleotides along each strand. Each base—A, C, T, or G—can be considered as a letter in a four-letter alphabet that spells out biological messages in the chemical structure of the DNA. organisms differ from one another because their respective DNA molecules have different nucleotide sequences and, consequently, carry different biological messages.

It was known well before the structure of DNA was determined that genes contain the instructions for producing proteins. The DNA messages must therefore somehow encode proteins. This relationship immediately makes the problem easier to understand, because of the chemical character of proteins. the properties of a protein, which are responsible for its biological function, are determined by its three-dimensional structure, and its structure is determined in turn by the linear sequence of the amino acids of which it is composed. The linear sequence of nucleotides in a gene must therefore somehow spell out the linear sequence of amino acids in a protein. The exact correspondence between the four-letter nucleotide alphabet of DNA and the twenty-letter amino acid alphabet of proteins—the genetic code—is not obvious from the DNA structure, and it took over a decade after the discovery of the double helix before it was worked out.

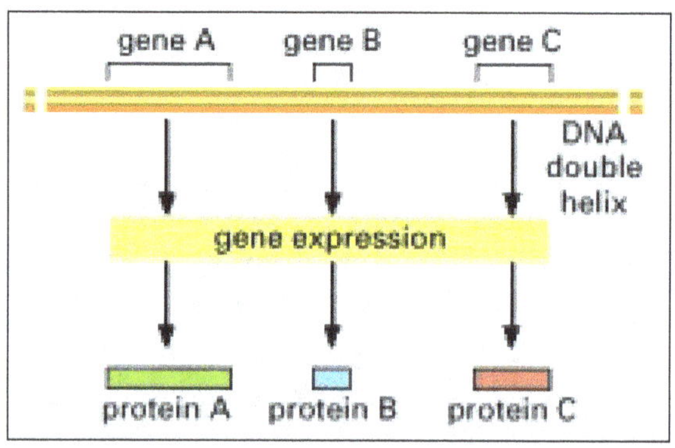

The relationship between genetic information carried in DNA and proteins.

The complete set of information in an organism's DNA is called its genome, and it carries the information for all the proteins the organism will ever synthesize. (The term genome is also used to describe the DNA that carries this information.) The amount of information contained in genomes is staggering: for example, a typical human cell contains 2 meters of DNA. Written out in the four-letter nucleotide alphabet, the nucleotide sequence of a very small human gene occupies a quarter of a page of text, while the complete sequence of nucleotides in the human genome would fill more than a thousand books the size of this one. In addition to other critical information, it carries the instructions for about 30,000 distinct proteins.

This gene carries the information for the amino acid sequence of one of the two types of subunits of the hemoglobin molecule, which carries oxygen in the blood. A different gene, the α-globin gene, carries the information for the other type of hemoglobin subunit (a hemoglobin molecule has four subunits, two of each type). Only one of the two strands of the DNA double helix containing the β-globin gene is shown; the other strand has the exact complementary sequence. By convention, a nucleotide sequence is written from its 5′

end to its 3′ end, and it should be read from left to right in successive lines down the page as though it were normal English text. The DNA sequences highlighted in yellow show the three regions of the gene that specify the amino sequence for the β-globin protein.

```
CCCTGTGGAGCCACACCCTAGGGTTGGCCA
ATCTACTCCCAGGAGCAGGGAGGGCAGGAG
CCAGGGCTGGGCATAAAAGTCAGGGCAGAG
CCATCTATTGCTTACATTTGCTTCTGACAC
AACTGTGTTCACTAGCAACTCAAACAGACA
CCATGGTGCACCTGACTCCTGAGGAGAAGT
CTGCCGTTACTGCCCTGTGGGGCAAGGTGA
ACGTGGATGAAGTTGGTGGTGAGGCCCTGG
GCAGGTTGGTATCAAGGTTACAAGACAGGT
TTAAGGAGACCAATAGAAACTGGGCATGTG
GAGACAGAGAAGACTCTTGGGTTTCTGATA
GGCACTGACTCTCTCTGCCTATTGGTCTAT
TTTCCCACCCTTAGGCTGCTGGTGGTCTAC
CCTTGGACCCAGAGGTTCTTTGAGTCCTTT
GGGGATCTGTCCACTCCTGATGCTGTTATG
GGCAACCCTAAGGTGAAGGCTCATGGCAAG
AAAGTGCTCGGTGCCTTTAGTGATGGCCTG
GCTCACCTGGACAACCTCAAGGGCACCTTT
GCCACACTGAGTGAGCTGCACTGTGACAAG
CTGCACGTGGATCCTGAGAACTTCAGGGTG
AGTCTATGGGACCCTTGATGTTTTCTTTCC
CCTTCTTTTCTATGGTTAAGTTCATGTCAT
AGGAAGGGGAGAAGTAACAGGGTACAGTTT
AGAATGGGAAACAGACGAATGATTGCATCA
GTGTGGAAGTCTCAGGATCGTTTTAGTTTC
TTTTATTTGCTGTTCATAACAATTGTTTTC
TTTTGTTTAATTCTTGCTTTCTTTTTTTTT
CTTCTCCGCAATTTTTACTATTATACTTAA
TGCCTTAACATTGTGTATAACAAAAGGAAA
TATCTCTGAGATACATTAAGTAACTTAAAA
AAAAACTTTACACAGTCTGCCTAGTACATT
ACTATTTGGAATATATGTGTGCTTATTTGC
ATATTCATAATCTCCCTACTTTATTTTCTT
TTATTTTTAATTGATACATAATCATTATAC
ATATTTATGGGTTAAAGTGTAATGTTTTAA
TATGTGTACACATATTGACCAAATCAGGGT
AATTTTGCATTTGTAATTTTAAAAAATGCT
TTCTTCTTTTAATATACTTTTTTGTTTATC
TTATTTCTAATACTTTCCCTAATCTCTTTC
TTTCAGGGCAATAATGATACAATGTATCAT
GCCTCTTTGCACCATTCTAAAGAATAACAG
TGATAATTTCTGGGTTAAGGCAATAGCAAT
ATTTCTGCATATAAATATTTCTGCATATAA
ATTGTAACTGATGTAAGAGGTTTCATATTG
CTAATAGCAGCTACAATCCAGCTACCATTC
TGCTTTTATTTTATGGTTGGGATAAGGCTG
GATTATTCTGAGTCCAAGCTAGGCCCTTTT
GCTAATCATGTTCATACCTCTTATCTTCCT
CCCACAGCTCCTGGGCAACGTGCTGGTCTG
TGTGCTGGCCCATCACTTTGGCAAAGAATT
CACCCCACCAGTGCAGGCTGCCTATCAGAA
AGTGGTGGCTGGTGTGGCTAATGCCCTGGC
CCACAAGTATCACTAAGCTCGCTTTCTTGC
TGTCCAATTTCTATTAAAGGTTCCTTTGTT
CCCTAAGTCCAACTACTAAACTGGGGGATA
TTATGAAGGGCCTTGAGCATCTGGATTCTG
CCTAATAAAAAACATTTATTTTCATTGCAA
TGATGTATTTAAATTATTTCTGAATATTTT
ACTAAAAAGGGAATGTGGGAGGTCAGTGCA
TTTAAAACATAAAGAAATGATGAGCTGTTC
AAACCTTGGGAAAATACACTATATCTTAAA
CTCCATGAAAGAAGGTGAGGCTGCAACCAG
CTAATGCACATTGGCAACAGCCCCTGATGC
CTATGCCTTATTCATCCCTCAGAAAAGGAT
TCTTGTAGAGGCTTGATTTGCAGGTTAAAG
TTTTGCTATGCTGTATTTTACATTACTTAT
TGTTTTAGCTGTCCTCATGAATGTCTTTTC
```

The nucleotide sequence of the human β-globin gene.

At each cell division, the cell must copy its genome to pass it to both daughter cells. The discovery of the structure of DNA also revealed the principle that makes this copying possible: because each strand of DNA contains a sequence of nucleotides that is exactly complementary to the nucleotide sequence of its partner strand, each strand can act as a template, or mold, for the synthesis of a new complementary strand. In other words, if we designate the two DNA strands as S and S′, strand S can serve as a template for making a new strand S′, while strand S′ can serve as a template for making a new strand S. Thus, the genetic information in DNA can be accurately copied by the beautifully simple process in which strand S separates from strand S′, and each separated strand then serves as a template for the production of a new complementary partner strand that is identical to its former partner.

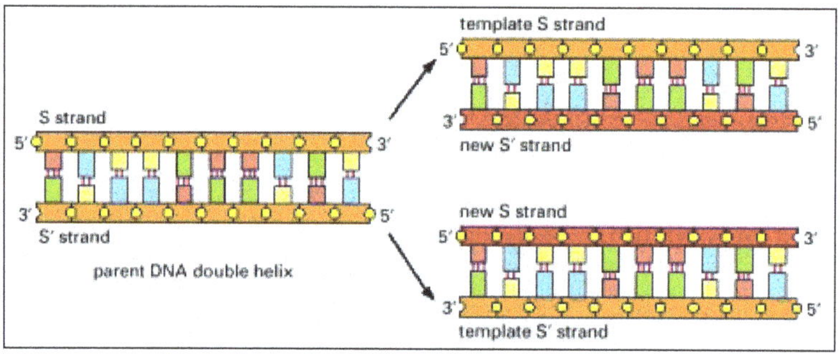

DNA as a template for its own duplication.

As the nucleotide A successfully pairs only with T, and G with C, each strand of DNA can specify the sequence of nucleotides in its complementary strand. In this way, double-helical DNA.

Physical and Chemical Properties of DNA

DNA is a long polymer made from repeating units called nucleotides. The DNA chain is 22 to 26 Ångströms wide (2.2 to 2.6 nanometres), and one nucleotide unit is 3.3 Ångstroms (0.33 nanometres) long. Although each individual repeating unit is very small, DNA polymers can be enormous molecules containing millions of nucleotides. For instance, the largest human chromosome, chromosome number 1, is 220 million base pairs long.

In living organisms, DNA does not usually exist as a single molecule, but instead as a tightly-associated pair of molecules. These two long strands entwine like vines, in the shape of a double helix. The nucleotide repeats contain both the segment of the backbone of the molecule, which holds the chain together, and a base, which interacts with the other DNA strand in the helix. In general, a base linked to a sugar is called a nucleoside and a base linked to a sugar and one or more phosphate groups is called a nucleotide. If multiple nucleotides are linked together, as in DNA, this polymer is referred to as a polynucleotide.

The backbone of the DNA strand is made from alternating phosphate and sugar residues. The sugar in DNA is 2-deoxyribose, which is a pentose (five-carbon) sugar. The sugars are joined together by phosphate groups that form phosphodiester bonds between the third and fifth carbon atoms of adjacent sugar rings. These asymmetric bonds mean a strand of DNA has a direction. In a double helix, the direction of the nucleotides in one strand is opposite to their direction in the other strand. This arrangement of DNA strands is called antiparallel. The asymmetric ends of DNA strands are referred to as the 5′ (five prime) and 3′ (three prime) ends. One of the major differences between DNA and RNA is the sugar, with 2-deoxyribose being replaced by the alternative pentose sugar ribose in RNA.

The DNA double helix is stabilized by hydrogen bonds between the bases attached to the two strands. The four bases found in DNA are adenine (abbreviated A), cytosine (C), guanine (G), and thymine (T). These four bases are shown below and are attached to the sugar/phosphate to form the complete nucleotide, as shown for adenosine monophosphate.

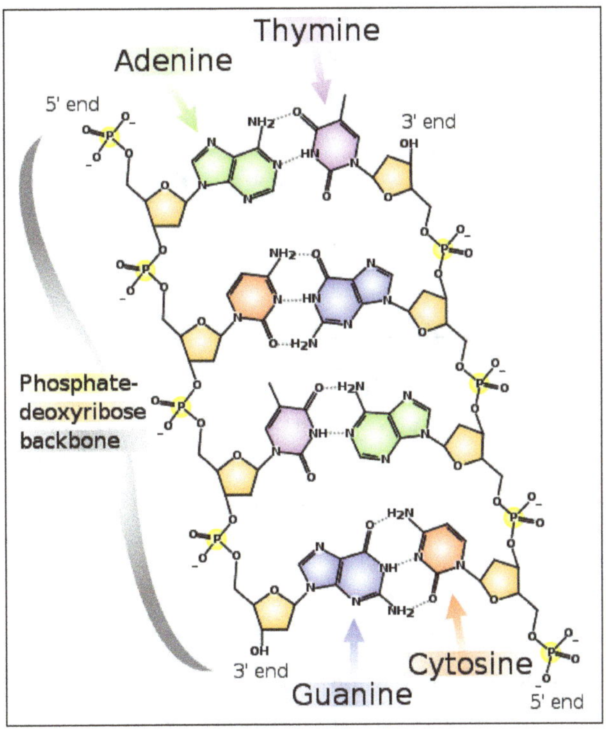

The chemical structure of DNA.

These bases are classified into two types; adenine and guanine are fused five- and six-membered heterocyclic compounds called purines, while cytosine and thymine are six-membered rings called pyrimidines. A fifth pyrimidine base, called uracil (U), usually takes the place of thymine in RNA and differs from thymine by lacking a methyl group on its ring. Uracil is not usually found in DNA, occurring only as a breakdown product of cytosine, but a very rare exception to this rule is a bacterial virus called PBS1 that contains uracil in its DNA. In contrast, following synthesis of certain RNA molecules, a significant number of the uracils are converted to thymines by the enzymatic addition of the missing methyl group. This occurs mostly on structural and enzymatic RNAs like transfer RNAs and ribosomal RNA.

Major and Minor Grooves

The double helix is a right-handed spiral. As the DNA strands wind around each other, they leave gaps between each set of phosphate backbones, revealing the sides of the bases inside. There are two of these grooves twisting around the surface of the double helix: one groove, the major groove, is 22 Å wide and the other, the minor groove, is 12 Å wide. The narrowness of the minor groove means that the edges of the bases are more accessible in the major groove. As a result, proteins like transcription factors that can bind to specific sequences in double-stranded DNA usually make contacts to the sides of the bases exposed in the major groove.

Base Pairing

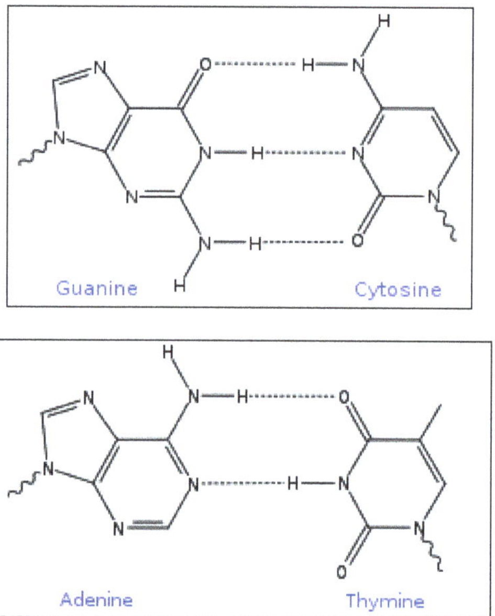

At top, a GC base pair with three hydrogen bonds. At the bottom, AT base pair with two hydrogen bonds. Hydrogen bonds are shown as dashed lines.

Each type of base on one strand forms a bond with just one type of base on the other strand. This is called complementary base pairing. Here, purines form hydrogen bonds to pyrimidines, with A bonding only to T, and C bonding only to G. This arrangement of two nucleotides binding together across the double helix is called a base pair. In a double helix, the two strands are also held together via forces generated by the hydrophobic effect and pi stacking, which are not influenced by the sequence of the DNA.

As hydrogen bonds are not covalent, they can be broken and rejoined relatively easily. The two strands of DNA in a double helix can therefore be pulled apart like a zipper, either by a mechanical force or high temperature. As a result of this complementarity, all the information in the double-stranded sequence of a DNA helix is duplicated on each strand, which is vital in DNA replication. Indeed, this reversible and specific interaction between complementary base pairs is critical for all the functions of DNA in living organisms.

The two types of base pairs form different numbers of hydrogen bonds, AT forming two hydrogen bonds, and GC forming three hydrogen bonds. The GC base pair is therefore stronger than the AT base pair. As a result, it is both the percentage of GC base pairs and the overall length of a DNA double helix that determine the strength of the association between the two strands of DNA. Long DNA helices with a high GC content have stronger-interacting strands, while short helices with high AT content have weaker-interacting strands. Parts of the DNA double helix that need to separate easily, such as the TATAAT Pribnow box in bacterial promoters, tend to have sequences with a high AT content, making the strands easier to pull apart. In the laboratory, the strength of this interaction can be measured by finding the temperature required to break the hydrogen bonds, their melting temperature (also called T_m value). When all the base pairs in a DNA double helix melt, the strands separate and exist in solution as two entirely independent molecules. These single-stranded DNA molecules have no single common shape, but some conformations are more stable than others.

Sense and Antisense

A DNA sequence is called “sense” if its sequence is the same as that of a messenger RNA copy that is translated into protein. The sequence on the opposite strand is complementary to the sense sequence and is therefore called the “antisense” sequence. Since RNA polymerases work by making a complementary copy of their templates, it is this antisense strand that is the template for producing the sense messenger RNA. Both sense and antisense sequences can exist on different parts of the same strand of DNA (that is, both strands contain both sense and antisense sequences).

In both prokaryotes and eukaryotes, antisense RNA sequences are produced, but the functions of these RNAs are not entirely clear. One proposal is that antisense RNAs are involved in regulating gene expression through RNA-RNA base pairing.

A few DNA sequences in prokaryotes and eukaryotes, and more in plasmids and viruses, blur the distinction made above between sense and antisense strands by having overlapping genes. In these cases, some DNA sequences do double duty, encoding one protein when read 5′ to 3′ along one strand, and a second protein when read in the opposite direction (still 5′ to 3′) along the other strand. In bacteria, this overlap may be involved in the regulation of gene transcription, while in viruses, overlapping genes increase the amount of information that can be encoded within the small viral genome.

Another way of reducing genome size is seen in some viruses that contain linear or circular single-stranded DNA as their genetic material.

Supercoiling

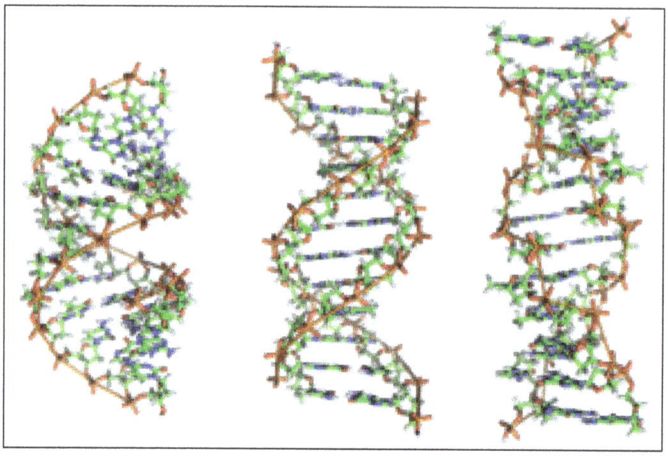

From left to right, the structures of A, B and Z DNA.

DNA can be twisted like a rope in a process called DNA supercoiling. With DNA in its "relaxed" state, a strand usually circles the axis of the double helix once every 10.4 base pairs, but if the DNA is twisted the strands become more tightly or more loosely wound. If the DNA is twisted in the direction of the helix, this is positive supercoiling, and the bases are held more tightly together. If they are twisted in the opposite direction, this is negative supercoiling, and the bases come apart more easily.

In nature, most DNA has slight negative supercoiling that is introduced by enzymes called topoisomerases. These enzymes are also needed to relieve the twisting stresses introduced into DNA strands during processes such as transcription and DNA replication.

Alternative Double-helical Structures

DNA exists in several possible conformations. The conformations so far identified are: A-DNA, B-DNA, C-DNA, D-DNA, E-DNA, H-DNA, L-DNA, P-DNA, and Z-DNA. However, only A-DNA, B-DNA, and Z-DNA have been observed in naturally occurring biological systems.

Which conformation DNA adopts depends on the sequence of the DNA, the amount and direction of supercoiling, chemical modifications of the bases, and also solution conditions, such as the concentration of metal ions and polyamines. Of these three conformations, the "B" form described above is most common under the conditions found in cells. The two alternative double-helical forms of DNA differ in their geometry and dimensions.

The A form is a wider right-handed spiral, with a shallow, wide minor groove and a narrower, deeper major groove. The A form occurs under non-physiological conditions in dehydrated samples of DNA, while in the cell it may be produced in hybrid pairings of DNA and RNA strands, as well as in enzyme-DNA complexes. Segments of DNA where the bases have been chemically-modified by methylation may undergo a larger change in conformation and adopt the Z form. Here, the strands turn about the helical axis in a left-handed spiral, the opposite of the more common B form. These unusual structures can be recognized by specific Z-DNA binding proteins and may be involved in the regulation of transcription.

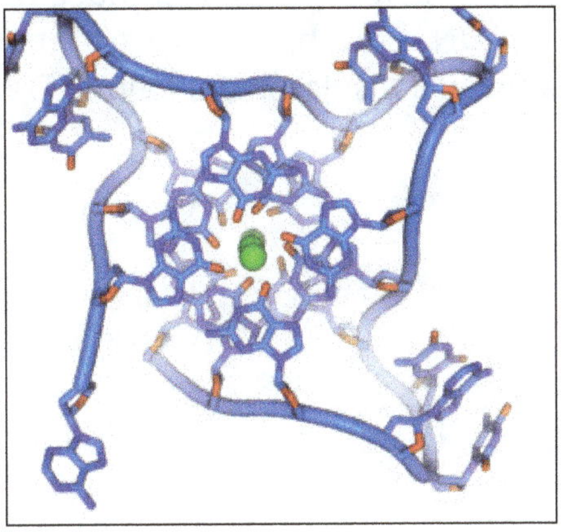

Structure of a DNA quadruplex formed by telomere repeats. The conformation of the DNA backbone diverges significantly from the typical helical structure.

Quadruplex Structures

At the ends of the linear chromosomes are specialized regions of DNA called telomeres. The main function of these regions is to allow the cell to replicate chromosome ends using the enzyme telomerase, as the enzymes that normally replicate DNA cannot copy the extreme 3′ ends of chromosomes. As a result, if a chromosome lacked telomeres it would become shorter each time it was replicated. These specialized chromosome caps also help protect the DNA ends from exonucleases and stop the DNA repair systems in the cell from treating them as damage to be corrected. In human cells, telomeres are usually lengths of single-stranded DNA containing several thousand repeats of a simple TTAGGG sequence.

These guanine-rich sequences may stabilize chromosome ends by forming very unusual structures of stacked sets of four-base units, rather than the usual base pairs found in other DNA molecules. Here, four guanine bases form a flat plate and these flat four-base units then stack on top of each other, to form a stable G-quadruplex structure. These structures are stabilized by hydrogen bonding between the edges of the bases and chelation of a metal ion in the centre of each four-base unit. The structure shown to the left is a top view of the quadruplex formed by a DNA sequence found in human telomere repeats. The single DNA strand forms a loop, with the sets of four bases stacking in a central quadruplex three plates deep. In the space at the center of the stacked bases are three chelated potassium ions. Other structures can also be formed, with the central set of four bases coming from either a single strand folded around the bases, or several different parallel strands, each contributing one base to the central structure.

In addition to these stacked structures, telomeres also form large loop structures called telomere loops, or T-loops. Here, the single-stranded DNA curls around in a long circle stabilized by telomere-binding proteins. At the very end of the T-loop, the single-stranded telomere DNA is held onto a region of double-stranded DNA by the telomere strand disrupting the double-helical DNA and base pairing to one of the two strands. This triple-stranded structure is called a displacement loop or D-loop.

2
Understanding RNA

The polymeric molecules which have vital biological roles in coding, decoding, regulation and expression of genes is known as RNA. The types of RNA include m-RNA, t-RNA, r-RNA and tmRNA. The topics elaborated in this chapter will help in gaining a better perspective about the functions, properties, and types of RNA.

Ribonucleic acid or RNA is a polymer or chain of nucleotide units, each comprising a nitrogenous base (adenine, cytosine, guanine, or uracil), a five-carbon sugar (ribose), and a phosphate group. The sugar and phosphate groups form the polymer's backbone, while the nitrogenous bases extending from the backbone provide RNA's distinctive properties.

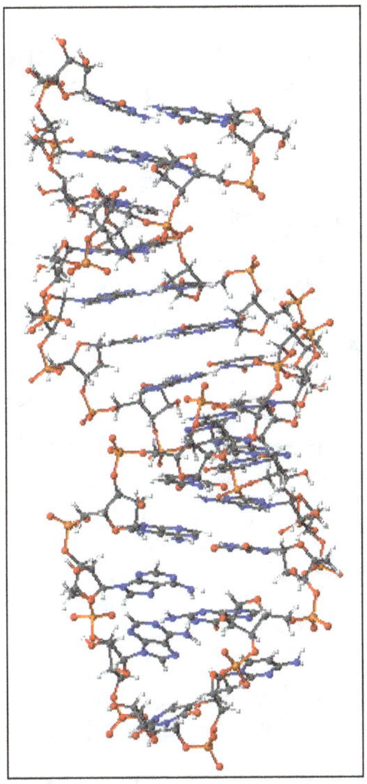

A hairpin loop from a pre-mRNA. Notice the single strand with its nitrogen-rich (blue) bases extending from its oxygen-rich (red) backbone.

In living cells, RNA in different configurations fulfills several important roles in the process of translating genetic information from deoxyribonucleic acid (DNA) into proteins. One type of RNA (messenger(m) RNA) acts as a messenger between DNA and the protein synthesis complexes known as ribosomes; a second type (ribosomal(r) RNA) forms vital portions of the structure of ribosomes; a third type (transfer(t) RNA) is an essential guide to deliver the appropriate protein building blocks, amino acids, to the ribosome; and other types of RNA, microRNAs (miRNAs) play a role in regulating gene expression, while small nuclear(sn) RNA helps with assuring that mRNA contains no nucleotide units that would lead to formation of a faulty protein. RNA also serves as a genetic blueprint for certain viruses, and some RNA molecules (called ribozymes) are also involved in the catalysis of biochemical reactions.

RNA is very similar to DNA, but differs in a few important structural details. RNA is usually single stranded, while DNA naturally seeks its stable form as a double stranded molecule. RNA nucleotides contain ribose while DNA nucleotides contain the closely related sugar deoxyribose. Furthermore, RNA uses the nucleotide uracil in its composition, instead of the thymine that is present in DNA. RNA is transcribed from DNA by enzymes called RNA polymerases and is generally further processed by other enzymes, some of them guided by non-coding RNAs.

Single-stranded RNA is similar to the protein polymer in its natural propensity to fold back and double up with itself in complex ways assuming a variety of biologically useful configurations.

The connectedness of living organisms can be seen in the ubiquitousness of RNA in living cells and in viruses throughout nature, and in the universal role of RNA in protein synthesis.

Chemical and Stereochemical Structure

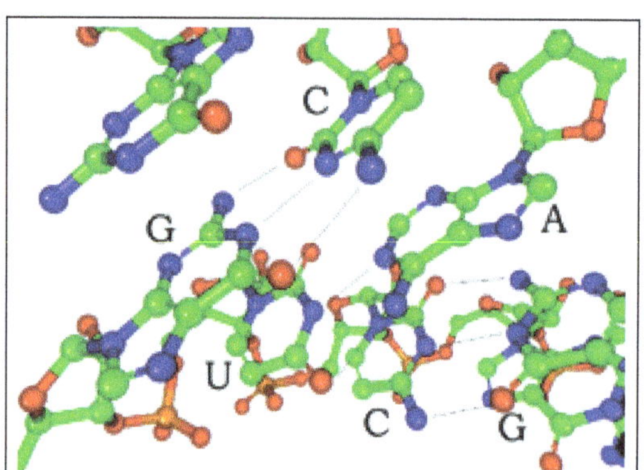

Base-pairing in a siRNA (small interfering RNA) segment, a double-stranded type of RNA. Hydrogen atoms are not shown.

RNA is a nucleic acid, a complex, high-molecular-weight macromolecule composed of nucleotide chains whose sequence of bases conveys genetic information.

A nucleotide is a chemical compound comprising three components: a nitrogen-containing base, a pentose (five-carbon) sugar, and one or more phosphate groups. The nitrogen-containing base of a nucleotide (also called the nucleobase) is typically a derivative of either purine or pyrimidine. The most common nucleotide bases are the purines adenine and guanine and the pyrimidines cytosine and thymine (or uracil in RNA).

Nucleic acids are polymers of repeating units (called monomers). Specifically, they often comprise long chains of nucleotide monomers connected by covalent chemical bonds. RNA molecules may comprise as few as 75 nucleotides or more than 5,000 nucleotides, while a DNA molecule may comprise more than 1,000,000 nucleotide units.

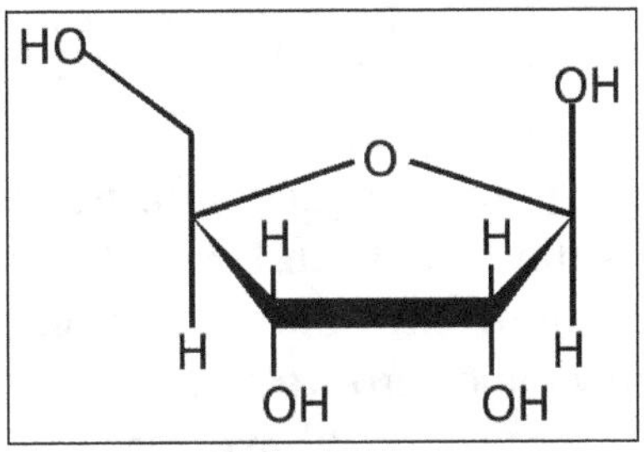

Ribose in acyclic form.

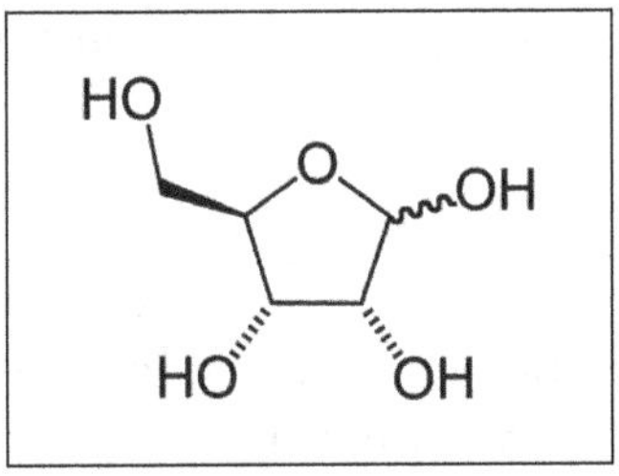

A conventional skeletal formula.

In RNA, the sugar component, ribose is a water-soluable, pentose sugar (monosaccharide with five carbon atoms). Ribose has the chemical formula $C_5H_{10}O_5$.

Ribose is an aldopentose, which means a pentose sugar with an aldehyde functional group in position one. An aldehyde group comprises a carbon atom bonded to a hydrogen atom and double-bonded to an oxygen atom (chemical formula O=CH-). Ribose forms a five-member ring with four carbon atoms and one oxygen. Hydroxyl (-OH) groups are attached to three of the carbons. The fourth carbon in the ring (one of the carbon atoms adjacent to the oxygen) has attached to it the fifth carbon atom and a hydroxyl group.

The RNA polymer features a ribose and phosphate backbone with one of four different nucleotide bases—adenine, guanine, cytosine, and uracil—attached to each ribose-phosphate unit.

There are also numerous modified bases and sugars found in RNA that serve many different roles. Pseudouridine (Ψ), in which the linkage between uracil and ribose is changed from a C–N bond to a C–C bond, and ribothymidine (T), are found in various places (most notably in the TΨC loop of tRNA). Another notable modified base is hypoxanthine (a deaminated guanine base whose nucleoside is called inosine). Inosine plays a key role in the Wobble Hypothesis of the genetic code. There are nearly 100 other naturally occurring modified nucleosides, of which pseudouridine and nucleosides with 2'-O-methylribose are by far the most common. The specific roles of many of these modifications in RNA are not fully understood. However, it is notable that in ribosomal RNA, many of the post-translational modifications occur in highly functional regions, such as the peptidyl transferase center and the subunit interface, implying that they are important for normal function.

The most important structural feature of RNA that distinguishes it from DNA is the presence of a hydroxyl group at the 2'-position of the ribose sugar. The presence of this functional group enforces the C3'-endo sugar conformation (as opposed to the C2'-endo conformation of the deoxyribose sugar in DNA) that causes the helix to adopt the A-form geometry rather than the B-form most commonly observed in DNA. This results in a very deep and narrow major groove and a shallow and wide minor groove. A second consequence of the presence of the 2'-hydroxyl group is that in conformationally flexible regions of an RNA molecule (that is, not involved in formation of a double helix), it can chemically attack the adjacent phosphodiester bond to cleave the backbone.

Synthesis

Synthesis of RNA is usually catalyzed by an enzyme, RNA polymerase, using DNA as a template. Initiation of synthesis begins with the binding of the enzyme to a promoter sequence in the DNA (usually found "upstream" of a gene). The DNA double helix is unwound by the helicase activity of the enzyme. The enzyme then progresses along the template strand in the 3' → 5' direction, synthesizing a complementary RNA molecule with elongation occurring in the 5' → 3' direction. The DNA sequence also dictates where termination of RNA synthesis will occur.

There are also a number of RNA-dependent RNA polymerases as well that use RNA as their template for synthesis of a new strand of RNA. For instance, a number of RNA viruses (such as poliovirus) use this type of enzyme to replicate their genetic material. Also, it is known that RNA-dependent RNA polymerases are required for the RNA interference pathway in many organisms.

Biological Roles

RNA's great variety of possible structures and chemical properties permits it to perform a much greater diversity of roles than in the cell than DNA. Three principal types of RNA are involved in protein synthesis:

- Messenger RNA (mRNA) serves as the template for the synthesis of a protein. It carries information from DNA to the ribosome.

- Transfer RNA (tRNA) is a small chain of nucleotides that transfers a specific amino acid to a growing polypeptide chain at the ribosomal site of synthesis. It pairs the amino acid to the appropriate three-nucleotide codon on the mRNA molecule.
- Ribosomal RNA (rRNA) molecules are extremely abundant and make up at least 80 percent of the RNA molecules found in a typical eukaryotic cell. In the cytoplasm, usually three or four rRNA molecules combine with many proteins to perform a structural and essential catalytic role, as components of the ribosome.

RNA also may serve as a catalyst for reactions and as a genetic blueprint, rather than DNA, in various viruses. Some RNA, including tRNA and rRNA, is non-coding in that it is not translated into proteins.

RNA Secondary Structures

The functional form of single stranded RNA molecules, just like proteins, frequently requires a specific tertiary structure. The scaffold for this structure is provided by secondary structural elements, which arise through the formation of hydrogen bonds within the interfolded molecule. This leads to several recognizable "domains" of secondary structure like hairpin loops, bulges, and internal loops. The secondary structure of RNA molecules can be predicted computationally by calculating the minimum free energies (MFE) structure for all different combinations of hydrogen bondings and domains. There has been a significant amount of research directed at the RNA structure prediction problem.

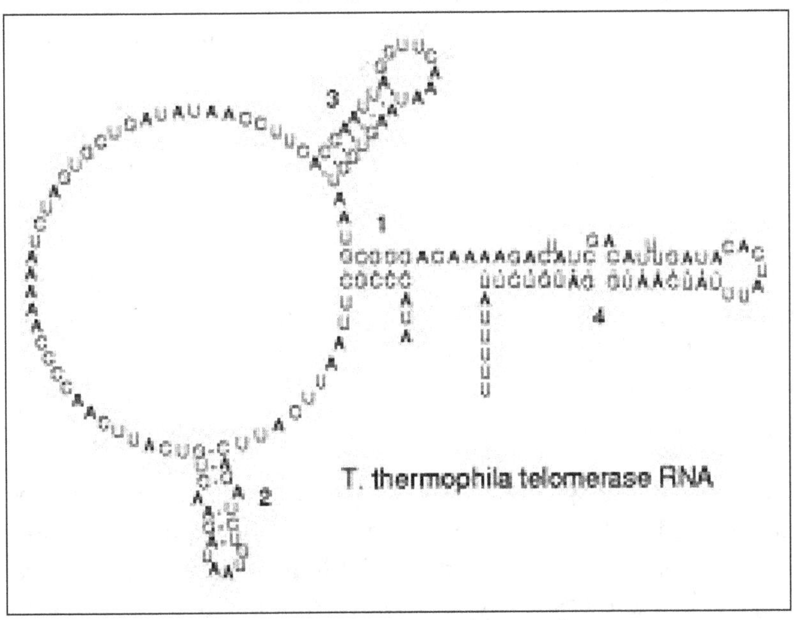

Secondary structure of an RNA from a telomerase.

Functions of RNA

Cells access the information stored in DNA by creating RNA to direct the synthesis of proteins through the process of translation. Proteins within a cell have many functions, including building cellular structures and serving as enzyme catalysts for cellular chemical reactions that give cells their specific characteristics. The three main types of RNA directly involved in protein synthesis are messenger RNA (mRNA), ribosomal RNA (rRNA), and transfer RNA (tRNA).

In 1961, French scientists François Jacob and Jacques Monod hypothesized the existence of an intermediary between DNA and its protein products, which they called messenger RNA. Evidence supporting their hypothesis was gathered soon afterwards showing that information from DNA is transmitted to the ribosome for protein synthesis using mRNA. If DNA serves as the complete library of cellular information, mRNA serves as a photocopy of specific information needed at a particular point in time that serves as the instructions to make a protein.

The mRNA carries the message from the DNA, which controls all of the cellular activities in a cell. If a cell requires a certain protein to be synthesized, the gene for this product is "turned on" and the mRNA is synthesized through the process of transcription. The mRNA then interacts with ribosomes and other cellular machinery to direct the synthesis of the protein it encodes during the process of translation. mRNA is relatively unstable and short-lived in the cell, especially in prokaryotic cells, ensuring that proteins are only made when needed.

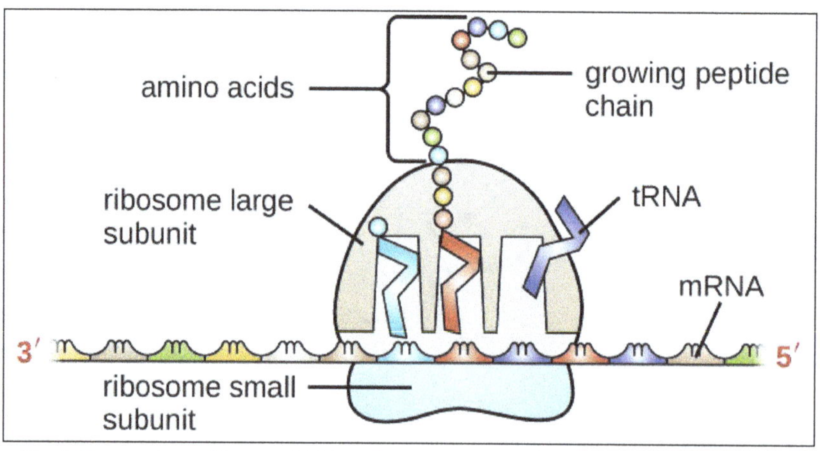

A generalized illustration of how mRNA and tRNA are used in protein synthesis within a cell.

rRNA and tRNA are stable types of RNA. In prokaryotes and eukaryotes, tRNA and rRNA are encoded in the DNA, then copied into long RNA molecules that are cut to release smaller fragments containing the individual mature RNA species. In eukaryotes, synthesis, cutting, and assembly of rRNA into ribosomes takes place in the nucleolus region of the nucleus, but these activities occur in the cytoplasm of prokaryotes.

Neither of these types of RNA carries instructions to direct the synthesis of a polypeptide, but they play other important roles in protein synthesis.

Ribosomes are composed of rRNA and protein. As its name suggests, rRNA is a major constituent of ribosomes, composing up to about 60% of the ribosome by mass and providing the location where the mRNA binds. The rRNA ensures the proper alignment of the mRNA, tRNA, and the ribosomes; the rRNA of the ribosome also has an enzymatic activity (peptidyl transferase) and catalyzes the formation of the peptide bonds between two aligned amino acids during protein synthesis. Although rRNA had long been thought to serve primarily a structural role, its catalytic role within the ribosome was proven in 2000. Scientists in the laboratories of Thomas Steitz and Peter Moore at Yale University were able to crystallize the ribosome structure from Haloarcula marismortui, a halophilic archaeon isolated from the Dead Sea. Because of the importance of this work, Steitz shared the 2009 Nobel Prize in Chemistry with other scientists who made significant contributions to the understanding of ribosome structure.

Transfer RNA is the third main type of RNA and one of the smallest, usually only 70–90 nucleotides long. It carries the correct amino acid to the site of protein synthesis in the ribosome. It is the base pairing between the tRNA and mRNA that allows for the correct amino acid to be inserted in the polypeptide chain being synthesized. Any mutations in the tRNA or rRNA can result in global problems for the cell because both are necessary for proper protein synthesis.

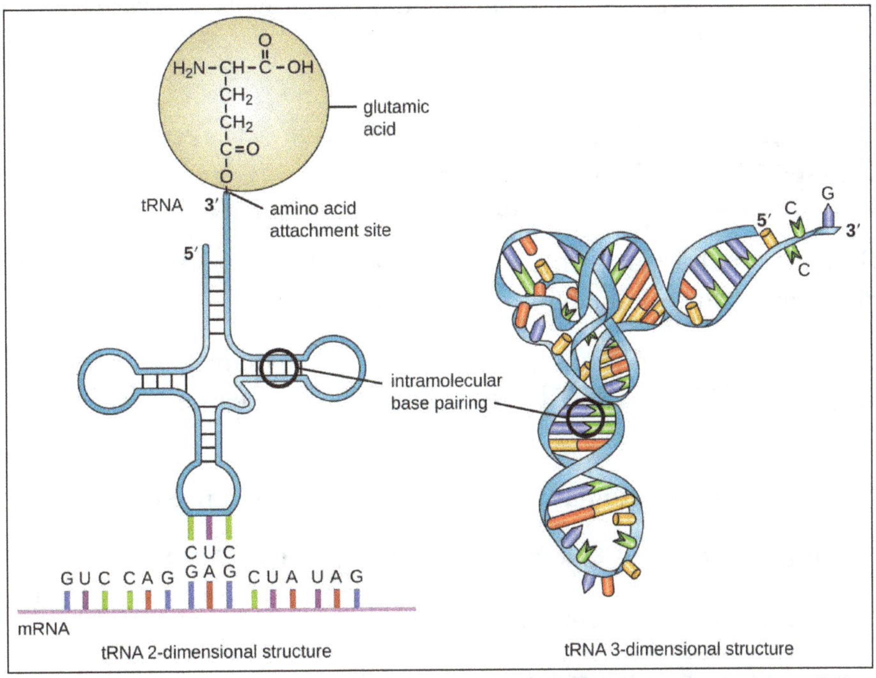

A tRNA molecule is a single-stranded molecule that exhibits significant intracellular base pairing, giving it its characteristic three-dimensional shape.

Table: Structure and Function of RNA.

	mRNA	rRNA	tRNA
Structure	Short, unstable, single-stranded RNA corresponding to a gene encoded within DNA	Longer, stable RNA molecules composing 60% of ribosome's mass	Short (70-90 nucleotides), stable RNA with extensive intramolecular base pairing; contains an amino acid binding site and an mRNA binding site
Function	Serves as intermediary between DNA and protein; used by ribosome to direct synthesis of protein it encodes	Ensures the proper alignment of mRNA, tRNA, and ribosome during protein synthesis; catalyzes peptide bond formation between amino acids	Carries the correct amino acid to the site of protein synthesis in the ribosome

RNA as Hereditary Information

Although RNA does not serve as the hereditary information in most cells, RNA does hold this function for many viruses that do not contain DNA. Thus, RNA clearly does have the additional capacity to serve as genetic information. Although RNA is typically single stranded within cells, there is significant diversity in viruses. Rhinoviruses, which cause the common cold; influenza viruses; and the Ebola virus are single-stranded RNA viruses. Rotaviruses, which cause severe gastroenteritis in children and other immunocompromised individuals, are examples of double-stranded RNA viruses. Because double-stranded RNA is uncommon in eukaryotic cells, its presence serves as an indicator of viral infection.

Properties of RNA

- RNA forms in the nucleolus, and then moves to specialized regions of the cytoplasm depending on the type of RNA formed.
- RNA, containing a ribose sugar, is more reactive than DNA and is not stable in alkaline conditions. RNA's larger helical grooves mean it is more easily subject to attack by enzymes.
- RNA strands are continually made, broken down and reused.
- RNA is more resistant to damage from UV light than DNA.
- RNA's mutation rate is relatively higher.
- Unusual bases may be present.
- The number of RNA may differ from cell to cell.

- Rate of renaturation after melting is quick.
- RNA is more versatile than DNA, capable of performing numerous, diverse tasks in an organism.

Types of RNA

m-RNA

Messenger RNA (mRNA) is a large family of RNA molecules that convey genetic information from DNA to the ribosome, where they specify the amino acid sequence of the protein products of gene expression. The RNA polymerase enzyme transcribes genes into primary transcript mRNA (known as pre-mRNA) leading to processed, mature mRNA. This mature mRNA is then translated into a polymer of amino acids: a protein, as summarized in the central dogma of molecular biology.

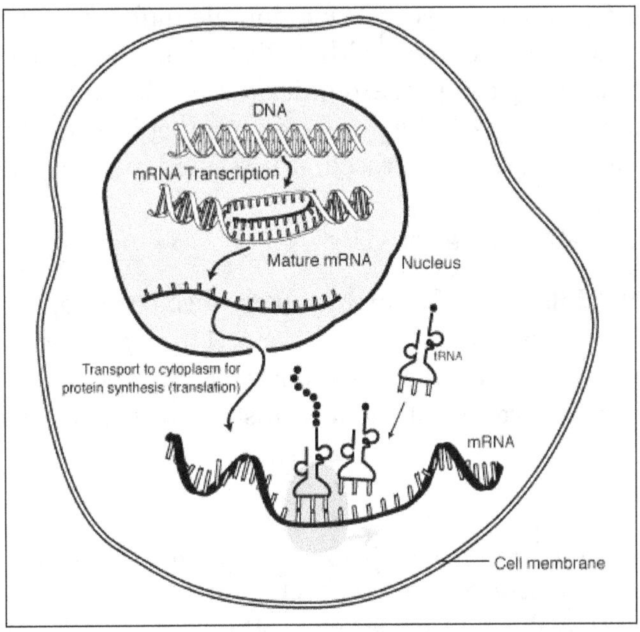

The "life cycle" of an mRNA in a eukaryotic cell. RNA is transcribed in the nucleus; after processing, it is transported to the cytoplasm and translated by the ribosome. Finally, the mRNA is degraded.

As in DNA, mRNA genetic information is in the sequence of nucleotides, which are arranged into codons consisting of three base pairs each. Each codon encodes for a specific amino acid, except the stop codons, which terminate protein synthesis. This process of translation of codons into amino acids requires two other types of RNA: Transfer RNA (tRNA), that mediates recognition of the codon and provides the corresponding amino acid, and ribosomal RNA (rRNA), that is the central component of the ribosome's protein-manufacturing machinery.

The existence of mRNA was first suggested by Jacques Monod and François Jacob, and subsequently discovered by Jacob, Sydney Brenner and Matthew Meselson at the California Institute of Technology in 1961.

It should not be confused with mitochondrial DNA.

Synthesis, Processing and Function

The brief existence of an mRNA molecule begins with transcription, and ultimately ends in degradation. During its life, an mRNA molecule may also be processed, edited, and transported prior to translation. Eukaryotic mRNA molecules often require extensive processing and transport, while prokaryotic mRNA molecules do not. A molecule of eukaryotic mRNA and the proteins surrounding it are together called a messenger RNP.

Transcription

Transcription is when RNA is made from DNA. During transcription, RNA polymerase makes a copy of a gene from the DNA to mRNA as needed. This process is similar in eukaryotes and prokaryotes. One notable difference, however, is that eukaryotic RNA polymerase associates with mRNA-processing enzymes during transcription so that processing can proceed quickly after the start of transcription. The short-lived, unprocessed or partially processed product is termed *precursor mRNA*, or *pre-mRNA*; once completely processed, it is termed *mature mRNA*.

Eukaryotic Pre-mRNA Processing

Processing of mRNA differs greatly among eukaryotes, bacteria, and archea. Non-eukaryotic mRNA is, in essence, mature upon transcription and requires no processing, except in rare cases. Eukaryotic pre-mRNA, however, requires several processing steps before its transport to the cytoplasm and its translation by the ribosome.

5' Cap Addition

A *5' cap* (also termed an RNA cap, an RNA 7-methylguanosine cap, or an RNA m^7G cap) is a modified guanine nucleotide that has been added to the "front" or 5' end of a eukaryotic messenger RNA shortly after the start of transcription. The 5' cap consists of a terminal 7-methylguanosine residue that is linked through a 5'-5'-triphosphate bond to the first transcribed nucleotide. Its presence is critical for recognition by the ribosome and protection from RNases.

Cap addition is coupled to transcription, and occurs co-transcriptionally, such that each influences the other. Shortly after the start of transcription, the 5' end of the mRNA being synthesized is bound by a cap-synthesizing complex associated with RNA polymerase. This enzymatic complex catalyzes the chemical reactions that are required for mRNA capping. Synthesis proceeds as a multi-step biochemical reaction.

Editing

In some instances, an mRNA will be edited, changing the nucleotide composition of that mRNA. An example in humans is the apolipopropein B mRNA, which is edited in some tissues, but not others. The editing creates an early stop codon, which, upon translation, produces a shorter protein.

Polyadenylation

Polyadenylation is the covalent linkage of a polyadenylyl moiety to a messenger RNA molecule. In eukaryotic organisms most messenger RNA (mRNA) molecules are polyadenylated at the 3' end, but recent studies have shown that short stretches of uridine (oligouridylation) are also common. The poly(A) tail and the protein bound to it aid in protecting mRNA from degradation by exonucleases. Polyadenylation is also important for transcription termination, export of the mRNA from the nucleus, and translation. mRNA can also be polyadenylated in prokaryotic organisms, where poly(A) tails act to facilitate, rather than impede, exonucleolytic degradation.

Polyadenylation occurs during and immediately after transcription of DNA into RNA. After transcription has been terminated, the mRNA chain is cleaved through the action of an endonuclease complex associated with RNA polymerase. After the mRNA has been cleaved, around 250 adenosine residues are added to the free 3' end at the cleavage site. This reaction is catalyzed by polyadenylate polymerase. Just as in alternative splicing, there can be more than one polyadenylation variant of an mRNA.

Polyadenylation site mutations also occur. The primary RNA transcript of a gene is cleaved at the poly-A addition site, and 100–200 A's are added to the 3' end of the RNA. If this site is altered, an abnormally long and unstable mRNA construct will be formed.

Transport

Another difference between eukaryotes and prokaryotes is mRNA transport. Because eukaryotic transcription and translation is compartmentally separated, eukaryotic mRNAs must be exported from the nucleus to the cytoplasm—a process that may be regulated by different signaling pathways. Mature mRNAs are recognized by their processed modifications and then exported through the nuclear pore by binding to the cap-binding proteins CBP20 and CBP80, as well as the transcription/export complex (TREX). Multiple mRNA export pathways have been identified in eukaryotes.

In spatially complex cells, some mRNAs are transported to particular subcellar destinations. In mature neurons, certain mRNA are transported from the soma to dendrites. One site of mRNA translation is at polyribosomes selectively localized beneath synapses. The mRNA for Arc/Arg3.1 is induced by synaptic activity and localizes selectively near active synapses based on signals generated by NMDA receptors. Other mRNAs also move into dendrites in response to external stimuli, such as β-actin mRNA. Upon export from the nucleus, actin mRNA associates with ZBP1 and the 40S subunit. The complex is bound by a motor protein and is transported to the target location (neurite extension) along the cytoskeleton. Eventually ZBP1 is phosphorylated by Src in order for translation to be initiated. In developing neurons, mRNAs are also transported into growing axons and especially growth cones. Many mRNAs are marked with so-called zip codes, which target their transport to a specific location.

Structure

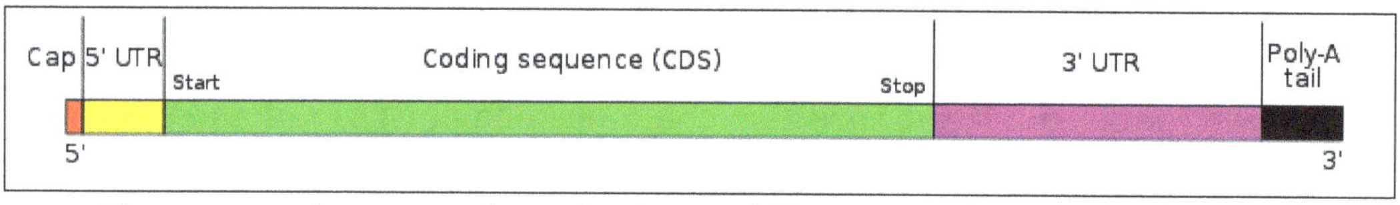

The structure of a mature eukaryotic mRNA. A fully processed mRNA includes a 5' cap, 5' UTR, coding region, 3' UTR, and poly(A) tail.

Coding Regions

Coding regions are composed of codons, which are decoded and translated (in eukaryotes usually into one and in prokaryotes usually into several) into proteins by the ribosome. Coding regions begin with the start codon and end with a stop codon. In general, the start codon is an AUG triplet and the stop codon is UAA, UAG, or UGA. The coding regions tend to be stabilised by internal base pairs, this impedes degradation. In addition to being protein-coding, portions of coding regions may serve as regulatory sequences in the pre-mRNA as exonic splicing enhancers or exonic splicing silencers.

Poly(A) Tail

The 3' poly(A) tail is a long sequence of adenine nucleotides (often several hundred) added to the 3' end of the pre-mRNA. This tail promotes export from the nucleus and translation, and protects the mRNA from degradation.

Monocistronic versus Polycistronic mRNA

An mRNA molecule is said to be monocistronic when it contains the genetic information to translate only a single protein chain (polypeptide). This is the case for most of the eukaryotic mRNAs. On the other hand, polycistronic mRNA carries several open reading frames (ORFs), each of which is translated into a polypeptide. These polypeptides usually have a related function (they often are the subunits composing a final complex protein) and their coding sequence is grouped and regulated together in a regulatory region, containing a promoter and an operator. Most of the mRNA found in bacteria and archaea is polycistronic, as is the human mitochondrial genome. Dicistronic or bicistronic mRNA encodes only two proteins.

mRNA Circularization

In eukaryotes, mRNA molecules form circular structures due to an interaction between the eIF4E and poly(A)-binding protein, which both bind to eIF4G, forming an mRNA-protein-mRNA bridge. Circularization is thought to promote cycling of ribosomes on the mRNA leading to time-efficient translation, and may also function to ensure only intact mRNA are translated (partially degraded mRNA characteristically have no m7G cap, or no poly-A tail).

Other mechanisms for circularization exist, particularly in virus mRNA. Poliovirus mRNA uses a cloverleaf section towards its 5' end to bind PCBP2, which binds poly(A)-binding protein, forming the familiar mRNA-protein-mRNA circle. Barley yellow dwarf virus has binding between mRNA segments on its 5' end and 3' end (called kissing stem loops), circularizing the mRNA without any proteins involved.

RNA virus genomes (the + strands of which are translated as mRNA) are also commonly circularized. Cite journal needed During genome replication the circularization acts to enhance genome replication speeds, cycling viral RNA-dependent RNA polymerase much the same as the ribosome is hypothesized to cycle.

Degradation

Different mRNAs within the same cell have distinct lifetimes (stabilities). In bacterial cells, individual mRNAs can survive from seconds to more than an hour. However, the lifetime averages between 1 and 3 minutes, making bacterial mRNA much less stable than eukaryotic mRNA. In mammalian cells, mRNA lifetimes range from several minutes to days. The greater the stability of an mRNA the more protein may be produced from that mRNA. The limited lifetime of mRNA enables a cell to alter protein synthesis rapidly in response to its changing needs. There are many mechanisms that lead to the destruction of an mRNA, some of which are described below.

Prokaryotic mRNA Degradation

In general, in prokaryotes the lifetime of mRNA is much shorter than in eukaryotes. Prokaryotes degrade messages by using a combination of ribonucleases, including endonucleases, 3' exonucleases, and 5' exonucleases. In some instances, small RNA molecules (sRNA) tens to hundreds of nucleotides long can stimulate the degradation of specific mRNAs by base-pairing with complementary sequences and facilitating ribonuclease cleavage by RNase III. It was recently shown that bacteria also have a sort of 5' cap consisting of a triphosphate on the 5' end. Removal of two of the phosphates leaves a 5' monophosphate, causing the message to be destroyed by the exonuclease RNase J, which degrades 5' to 3'.

Eukaryotic mRNA Turnover

Inside eukaryotic cells, there is a balance between the processes of translation and mRNA decay. Messages that are being actively translated are bound by ribosomes, the eukaryotic initiation factors eIF-4E and eIF-4G, and poly(A)-binding protein. eIF-4E and eIF-4G block the decapping enzyme (DCP2), and poly(A)-binding protein blocks the exosome complex, protecting the ends of the message. The balance between translation and decay is reflected in the size and abundance of cytoplasmic structures known as P-bodies. The poly(A) tail of the mRNA is shortened by specialized exonucleases that are targeted to specific messenger RNAs by a combination of cis-regulatory sequences on the RNA and trans-acting RNA-binding proteins. Poly(A) tail removal is thought to disrupt the circular structure of the message and destabilize the cap binding complex. The message is then subject to degradation by either the exosome complex or the decapping complex. In this way, translationally inactive messages can be destroyed quickly, while active messages remain intact. The mechanism by which translation stops and the message is handed-off to decay complexes is not understood in detail.

AU-rich Element Decay

The presence of AU-rich elements in some mammalian mRNAs tends to destabilize those transcripts through the action of cellular proteins that bind these sequences and stimulate poly(A) tail removal. Loss of the poly(A) tail is thought to promote mRNA degradation by facilitating attack by both the exosome complex and the decapping complex. Rapid mRNA degradation via AU-rich elements is a critical mechanism for preventing the overproduction of potent cytokines such as tumor necrosis factor (TNF) and granulocyte-macrophage colony stimulating factor (GM-CSF). AU-rich elements also regulate the biosynthesis of proto-oncogenic transcription factors like c-Jun and c-Fos.

Nonsense Mediated Decay

Eukaryotic messages are subject to surveillance by nonsense mediated decay (NMD), which checks for the presence of premature stop codons (nonsense codons) in the message. These can arise via incomplete splicing, V(D)J recombination in the adaptive immune system, mutations in DNA, transcription errors, leaky scanning by the ribosome causing a frame shift, and other causes. Detection of a premature stop codon triggers mRNA degradation by 5' decapping, 3' poly(A) tail removal, or endonucleolytic cleavage.

Small Interfering RNA (siRNA)

In metazoans, small interfering RNAs (siRNAs) processed by Dicer are incorporated into a complex known as the RNA-induced silencing complex or RISC. This complex contains an endonuclease that cleaves perfectly complementary messages to which the siRNA binds. The resulting mRNA fragments are then destroyed by exonucleases. siRNA is commonly used in laboratories to block the function of genes in cell culture. It is thought to be part of the innate immune system as a defense against double-stranded RNA viruses.

MicroRNA (miRNA)

MicroRNAs (miRNAs) are small RNAs that typically are partially complementary to sequences in metazoan messenger RNAs. Binding of a miRNA to a message can repress translation of that message and accelerate poly(A) tail removal, thereby hastening mRNA degradation. The mechanism of action of miRNAs is the subject of active research.

Other Decay Mechanisms

There are other ways by which messages can be degraded, including non-stop decay and silencing by Piwi-interacting RNA (piRNA), among others.

mRNA-based Therapeutics

Full length mRNA molecules have been proposed as therapeutics since the beginning of the biotech era but there was little traction until the 2010s, when Moderna Therapeutics was founded and managed to raise almost a billion dollars in venture funding in its first three years.

Theoretically, the administered mRNA sequence can cause a cell to make a protein, which in turn could directly treat a disease or could function as a vaccine; more indirectly the protein could drive an endogenous stem cell to differentiate in a desired way.

The primary challenges of RNA therapy center on delivering the RNA to directed cells, more even than determining what sequence to deliver. Naked RNA sequences will naturally degrade after preparation; they may trigger the body's immune system to attack them as an invader; and they are impermeable to the cell membrane. Once within the cell, they must then leave the cell's transport mechanism to take action within the cytoplasm, which houses the ribosomes that direct manufacture of proteins.

tRNA

A transfer RNA (abbreviated tRNA and formerly referred to as sRNA, for soluble RNA) is an adaptor molecule composed of RNA, typically 76 to 90 nucleotides in length, that serves as the physical link between the mRNA and the amino acid sequence of proteins. tRNA does this by carrying an amino acid to the protein synthetic machinery of a cell (ribosome) as directed by a 3-nucleotide sequence (codon) in a messenger RNA (mRNA). As such, tRNAs are a necessary component of translation, the biological synthesis of new proteins in accordance with the genetic code.

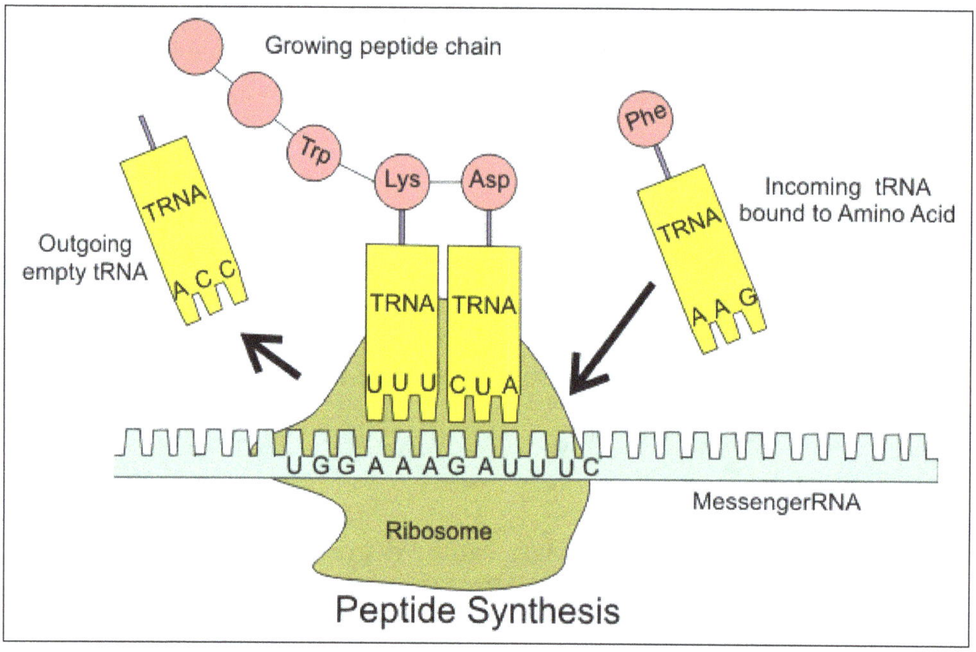

The interaction of tRNA and mRNA in protein synthesis.

While the specific nucleotide sequence of an mRNA specifies which amino acids are incorporated into the protein product of the gene from which the mRNA is transcribed, the role of tRNA is to specify which sequence from the genetic code corresponds to which amino acid. The mRNA encodes a protein as a series of contiguous codons, each of which is recognized by a particular tRNA. One end of the tRNA matches the genetic code in a three-nucleotide sequence called the anticodon. The anticodon forms three complementary base pairs with a codon in mRNA during protein biosynthesis. On the other end of the tRNA is a covalent attachment to the amino acid that corresponds to the anticodon sequence. Each type of tRNA molecule can be attached to only one type of amino acid, so each organism has many types of tRNA. Because the genetic code contains multiple codons that specify the same amino acid, there are several tRNA molecules bearing different anticodons which carry the same amino acid. The covalent attachment to the tRNA 3' end is catalyzed by enzymes called aminoacyl tRNA synthetases. During protein synthesis, tRNAs with attached amino acids are delivered to the ribosome by proteins called elongation factors, which aid in association of the tRNA with the ribosome, synthesis of the new polypeptide, and translocation (movement) of the ribosome along the mRNA. If the tRNA's anticodon matches the mRNA, another tRNA already bound to the ribosome transfers the growing polypeptide chain from its 3' end to the amino acid attached to the 3' end of the newly delivered tRNA, a reaction catalyzed by the ribosome. A large number of the individual nucleotides in a tRNA molecule may be chemically modified, often by methylation or deamidation. These unusual bases sometimes affect the tRNA's interaction with ribosomes and sometimes occur in the anticodon to alter base-pairing properties.

Structure

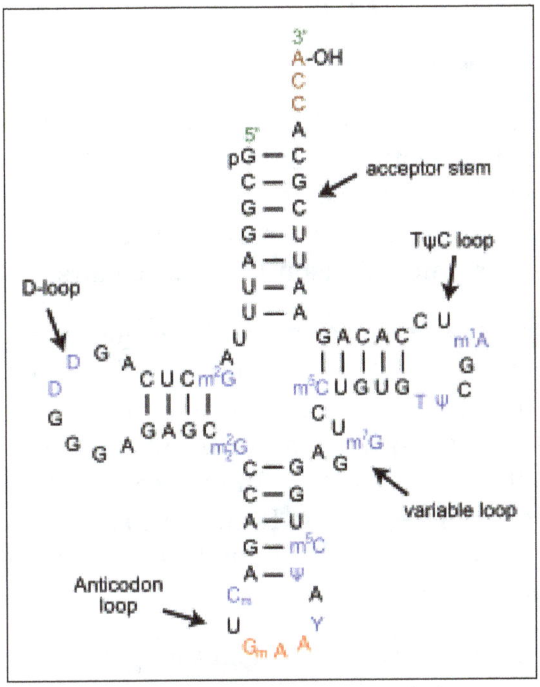

Secondary *cloverleaf structure* of tRNA^{Phe} from yeast.

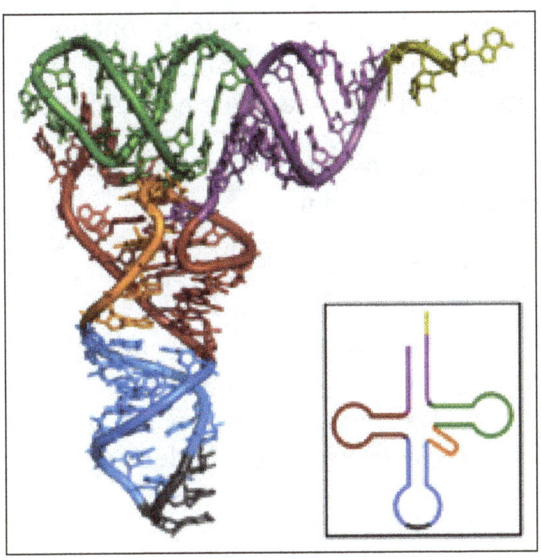

Tertiary structure of tRNA: CCA tail in yellow, Acceptor stem in purple, Variable loop in orange, D arm in red, Anticodon arm in blue with Anticodon in black, T arm in green.

The structure of tRNA can be decomposed into its primary structure, its secondary structure (usually visualized as the *cloverleaf structure*), and its tertiary structure (all tRNAs have a similar L-shaped 3D structure that allows them to fit into the P and A sites of the ribosome). The cloverleaf structure becomes the 3D L-shaped structure through coaxial stacking of the helices, which is a common RNA tertiary structure motif. The lengths of each arm, as well as the loop 'diameter', in a tRNA molecule vary from species to species. The tRNA structure consists of the following:

- A 5'-terminal phosphate group.
- The acceptor stem is a 7- to 9-base pair (bp) stem made by the base pairing of the 5'-terminal nucleotide with the 3'-terminal nucleotide (which contains the CCA 3'-terminal group used to attach the amino acid). In general, such 3'-terminal tRNA-like structures are referred to as 'genomic tags'. The acceptor stem may contain non-Watson-Crick base pairs.
- The CCA tail is a cytosine-cytosine-adenine sequence at the 3' end of the tRNA molecule. The amino acid loaded onto the tRNA by aminoacyl tRNA synthetases, to form aminoacyl-tRNA, is covalently bonded to the 3'-hydroxyl group on the CCA tail. This sequence is important for the recognition of tRNA by enzymes and critical in translation. In prokaryotes, the CCA sequence is transcribed in some tRNA sequences. In most prokaryotic tRNAs and eukaryotic tRNAs, the CCA sequence is added during processing and therefore does not appear in the tRNA gene.
- The D arm is a 4- to 6-bp stem ending in a loop that often contains dihydrouridine.
- The anticodon arm is a 5-bp stem whose loop contains the anticodon. The tRNA 5'-to-3' primary structure contains the anticodon but in reverse order, since 3'-to-5' directionality is required to read the mRNA from 5'-to-3'.
- The T arm is a 4- to 5- bp stem containing the sequence TΨC where Ψ is pseudouridine, a modified uridine.
- Bases that have been modified, especially by methylation (e.g. tRNA (guanine-N7-)-methyltransferase), occur in several positions throughout the tRNA. The first anticodon base, or wobble-position, is sometimes modified to inosine (derived from adenine), queuosine (derived from guanine), uridine-5-oxyacetic acid (derived from uracil), 5-methylaminomethyl-2-thiouridine (derived from uracil), or lysidine (derived from cytosine).

Anticodon

An anticodon is a unit made up of three nucleotides that correspond to the three bases of the codon on the mRNA. Each tRNA contains a distinct anticodon triplet sequence that can form 3 complementary base pairs to one or more codons for an amino acid. Some anticodons can pair with more than one codon due to a phenomenon known as wobble base pairing. Frequently, the first nucleotide of the anticodon is one not found on mRNA: inosine, which can hydrogen bond to more than one base in the corresponding codon position.:29.3.9 In the genetic code, it is common for a single amino acid to be specified by all four third-position possibilities, or at least by both pyrimidines and purines; for example, the amino acid glycine is coded for by the codon sequences GGT, GGC, GGA, and GGG. Other modified nucleotides may also appear at the first anticodon position—sometimes known as the "wobble position"—resulting in subtle changes to the genetic code, as for example in mitochondria. Per cell, 61 types of tRNA would be required to provide a one-to-one correspondence between tRNA molecules and codons that specify amino acids, as there are 61 sense codons of the standard genetic code. However, many cells contain fewer than 61 types of tRNAs because the wobble base is capable of binding to several, though not necessarily all, of the codons that specify a particular amino acid. A minimum of 31 tRNAs are required to translate, unambiguously, all 61 sense codons; the maximum observed is 41.

Aminoacylation

Aminoacylation is the process of adding an aminoacyl group to a compound. It covalently links an amino acid to the CCA 3' end of a tRNA molecule. Each tRNA is aminoacylated (or *charged*) with a specific amino acid by an aminoacyl tRNA synthetase. There is normally a single aminoacyl tRNA synthetase for each amino acid, despite the fact that there can be more than one tRNA, and more than one anticodon for an amino acid. Recognition of the appropriate tRNA by the synthetases is not mediated solely by the anticodon, and the acceptor stem often plays a prominent role. Reaction:

- Amino acid + ATP → aminoacyl-AMP + PPi.
- Aminoacyl-AMP + tRNA → aminoacyl-tRNA + AMP.

Certain organisms can have one or more aminoacyl tRNA synthetases missing. This leads to charging of the tRNA by a chemically related amino acid, and by use of an enzyme or enzymes, the tRNA is modified to be correctly charged. For example, *Helicobacter pylori* has glutaminyl tRNA synthetase missing. Thus, glutamate tRNA synthetase charges tRNA-glutamine(tRNA-Gln) with glutamate. An amidotransferase then converts the acid side chain of the glutamate to the amide, forming the correctly charged gln-tRNA-Gln.

Binding to Ribosome

The ribosome has three binding sites for tRNA molecules that span the space between the two ribosomal subunits: the A (aminoacyl), P (peptidyl), and E (exit) sites. In addition, the ribosome has two other sites for tRNA binding that are used during mRNA decoding or during the initiation of protein synthesis. These are the T site (named elongation factor Tu) and I site (initiation). By convention, the tRNA binding sites are denoted with the site on the small ribosomal subunit listed first and the site on the large ribosomal subunit listed second. For example, the A site is often written A/A, the P site, P/P, and the E site, E/E. The binding proteins like L27, L2, L14, L15, L16 at the A- and P- sites have been determined by affinity labeling by A.P. Czernilofsky et al.

Once translation initiation is complete, the first aminoacyl tRNA is located in the P/P site, ready for the elongation cycle described below. During translation elongation, tRNA first binds to the ribosome as part of a complex with elongation factor Tu (EF-Tu) or its eukaryotic (eEF-1) or archaeal counterpart. This initial tRNA binding site is called the A/T site. In the A/T site, the A-site half resides in the small ribosomal subunit where the mRNA decoding site is located. The mRNA decoding site is where the mRNA codon is read out during translation. The T-site half resides mainly on the large ribosomal subunit where EF-Tu or eEF-1 interacts with the ribosome. Once mRNA decoding is complete, the aminoacyl-tRNA is bound in the A/A site and is ready for the next peptide bond to be formed to its attached amino acid. The peptidyl-tRNA, which transfers the growing polypeptide to the aminoacyl-tRNA bound in the A/A site, is bound in the P/P site. Once the peptide bond is formed, the tRNA in the P/P site is deacylated, or has a free 3' end, and the tRNA in the A/A site carries the growing polypeptide chain. To allow for the next elongation cycle, the tRNAs then move through hybrid A/P and P/E binding sites, before completing the cycle and residing in the P/P and E/E sites. Once the A/A and P/P tRNAs have moved to the P/P and E/E sites, the mRNA has also moved over by one codon and the A/T site is vacant, ready for the next round of mRNA decoding. The tRNA bound in the E/E site then leaves the ribosome.

The P/I site is actually the first to bind to aminoacyl tRNA, which is delivered by an initiation factor called IF2 in bacteria. However, the existence of the P/I site in eukaryotic or archaeal ribosomes has not yet been confirmed. The P-site protein L27 has been determined by affinity labeling by E. Collatz and A.P. Czernilofsky.

tRNA Genes

Organisms vary in the number of tRNA genes in their genome. For example, the nematode worm *C. elegans*, a commonly used model organism in genetics studies, has 29,647 genes in its nuclear genome, of which 620 code for tRNA. The budding yeast *Saccharomyces cerevisiae* has 275 tRNA genes in its genome.

In the human genome, which, according to January 2013 estimates, has about 20,848 protein coding genes in total, there are 497 nuclear genes encoding cytoplasmic tRNA molecules, and 324 tRNA-derived pseudogenes—tRNA genes thought to be no longer functional (although pseudo tRNAs have been shown to be involved in antibiotic resistance in bacteria). Regions in nuclear chromosomes, very similar in sequence to mitochondrial tRNA genes, have also been identified (tRNA-lookalikes). These tRNA-lookalikes are also considered part of the nuclear mitochondrial DNA (genes transferred from the mitochondria to the nucleus).

As with all eukaryotes, there are 22 mitochondrial tRNA genes in humans. Mutations in some of these genes have been associated with severe diseases like the MELAS syndrome.

Cytoplasmic tRNA genes can be grouped into 49 families according to their anticodon features. These genes are found on all chromosomes, except the 22 and Y chromosome. High clustering on 6p is observed (140 tRNA genes), as well on 1 chromosome.

The HGNC, in collaboration with the Genomic tRNA Database (GtRNAdb) and experts in the field, has approved unique names for human genes that encode tRNAs.

Evolution

The top half of tRNA (consisting of the T arm and the acceptor stem with 5'-terminal phosphate group and 3'-terminal CCA group) and the bottom half (consisting of the D arm and the anticodon arm) are independent units in structure as well as in function. The top half may have evolved first including the 3'-terminal genomic tag which originally may have marked tRNA-like molecules for replication in early RNA world. The bottom half may have evolved later as an expansion, e.g. as protein synthesis started in RNA world and turned it into a ribonucleoprotein world (RNP world). This proposed scenario is called genomic tag hypothesis. In fact, tRNA and tRNA-like aggregates have an important catalytic influence (i. e. as ribozymes) on replication still today. These roles may be regarded as 'molecular (or chemical) fossils' of RNA world.

Genomic tRNA content is a differentiating feature of genomes among biological domains of life: Archaea present the simplest situation in terms of genomic tRNA content with a uniform number of gene copies, Bacteria have an intermediate situation and Eukarya present the most complex situation. Eukarya present not only more tRNA gene content than the other two kingdoms but also a high variation in gene copy number among different isoacceptors, and this complexity seem to be due to duplications of tRNA genes and changes in anticodon specificity.

Evolution of the tRNA gene copy number across different species has been linked to the appearance of specific tRNA modification enzymes (uridine methyltransferases in Bacteria, and adenosine deaminases in Eukarya), which increase the decoding capacity of a given tRNA. As an example, tRNAAla encodes four different tRNA isoacceptors (AGC, UGC, GGC and CGC). In Eukarya, AGC isoacceptors are extremely enriched in gene copy number in comparison to the rest of isoacceptors, and this has been correlated with its A-to-I modification of its wobble base. This same trend has been shown for most amino acids of eukaryal species. Indeed, the effect of these two tRNA modifications is also seen in codon usage bias. Highly expressed genes seem to be enriched in codons that are exclusively using codons that will be decoded by these modified tRNAs, which suggests a possible role of these codons—and consequently of these tRNA modifications—in translation efficiency.

tRNA-derived Fragments

tRNA-derived fragments (or tRFs) are short molecules that emerge after cleavage of the mature tRNAs or the precursor transcript. Both cytoplasmic and mitochondrial tRNAs can produce fragments. There are at least four structural types of tRFs believed to originate from mature tRNAs, including the relatively long tRNA halves and short 5'-tRFs, 3'-tRFs and i-tRFs. The precursor tRNA can be cleaved to produce molecules from the 5' leader or 3' trail sequences. Cleavage enzymes include Angiogenin, Dicer, RNase Z and RNase P. Especially in the case of Angiogenin, the tRFs have a characteristically unusual cyclic phosphate at their 3' end and a hydroxyl group at the 5' end. tRFs appear to play a role in RNA interference, specifically in the surpression of retroviruses and retrotransposons that use tRNA as a primer for replication. Half-tRNAs cleaved by angiogenin are also known as tiRNAs. The biogensis of smaller fragments, including those that function as piRNAs, are less understood.

tRFs have multiple dependencies and roles; such as exhibiting significant changes between sexes, among races and disease status. Functionally, they can be loaded on Ago and act through RNAi pathways, participate in the formation of stress granules, displace mRNAs from RNA-binding proteins or inhibit translation. At the system or the organismal level, the four types of tRFs have a diverse spectrum of activities. Functionally, tRFs are associated with viral infection, cancer, cell proliferation and also with epigenetic transgenerational regulation of metabolism.

tRFs are not restricted to humans and have been shown to exist in multiple organisms.

Two online tools are available for those wishing to learn more about tRFs: the framework for the interactive exploration of mitochondrial and nuclear tRNA fragments (MINTbase) and the relational database of Transfer RNA related Fragments (tRFdb). MINTbase also provides a naming scheme for the naming of tRFs called tRF-license plates (or MINTcodes) that is genome independent; the scheme compresses an RNA sequence into a shorter string.

Engineered tRNAs

Artificial suppressor elongator tRNAs are used to incorporate unnatural amino acids at nonsense codons placed in the coding sequence of a gene. Engineered initiator tRNAs ($tRNA^{fMet2}$ with CUA anticodon encoded by metY gene) have been used to initiate translation at the amber stop codon UAG. This type of engineered tRNA is called a nonsense suppressor tRNA because it suppresses the translation stop signal that normally occurs at UAG codons. The amber initiator tRNA inserts methionine and glutamine at UAG codons preceded by a strong Shine-Dalgarno sequence. An investigation of the amber initiator tRNA showed that it was orthogonal to the regular AUG start codon showing no detectable off-target translation initiation events in a genomically recoded *E. coli* strain.

tRNA Biogenesis

In eukaryotic cells, tRNAs are transcribed by RNA polymerase III as pre-tRNAs in the nucleus. RNA polymerase III recognizes two highly conserved downstream promoter sequences: the 5' intragenic control region (5'-ICR, D-control region, or A box), and the 3'-ICR (T-control region or B box) inside tRNA genes. The first promoter begins at +8 of mature tRNAs and the second promoter is located 30–60 nucleotides downstream of the first promoter. The transcription terminates after a stretch of four or more thymidines.

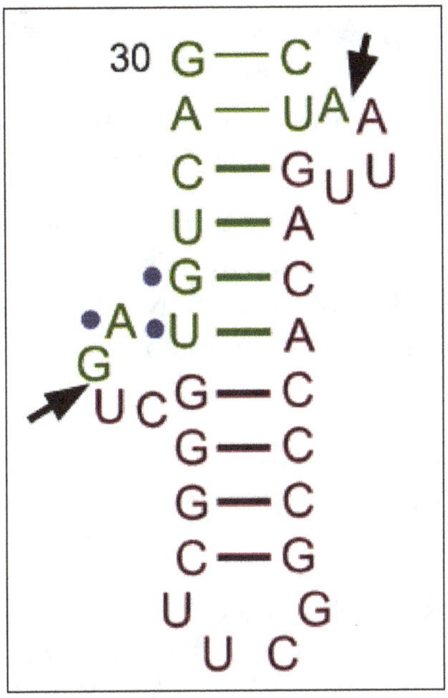

Bulge-helix-bulge motive of tRNA intron.

Pre-tRNAs undergo extensive modifications inside the nucleus. Some pre-tRNAs contain introns that are spliced, or cut, to form the functional tRNA molecule; in bacteria these self-splice, whereas in eukaryotes and archaea they are removed by tRNA-splicing endonucleases. Eukaryotic pre-tRNA contains bulge-helix-bulge (BHB) structure motif that is important for recognition and precise splicing of tRNA intron by endonucleases. This motif position and structure are evolutionary conserved. However, some organisms, such as unicellular algae have a non-canonical position of BHB-motif as well as 5'- and 3'-ends of the spliced intron sequence. The 5' sequence is removed by RNase P, whereas the 3' end is removed by the tRNase Z enzyme. A notable exception is in the archaeon *Nanoarchaeum equitans,* which does not possess an RNase P enzyme and has a promoter placed such that transcription starts at the 5' end of the mature tRNA. The non-templated 3' CCA tail is added by a nucleotidyl transferase. Before tRNAs are exported into the cytoplasm by Los1/Xpo-t, tRNAs are aminoacylated. The order of the processing events is not conserved. For example, in yeast, the splicing is not carried out in the nucleus but at the cytoplasmic side of mitochondrial membranes.

rRNA

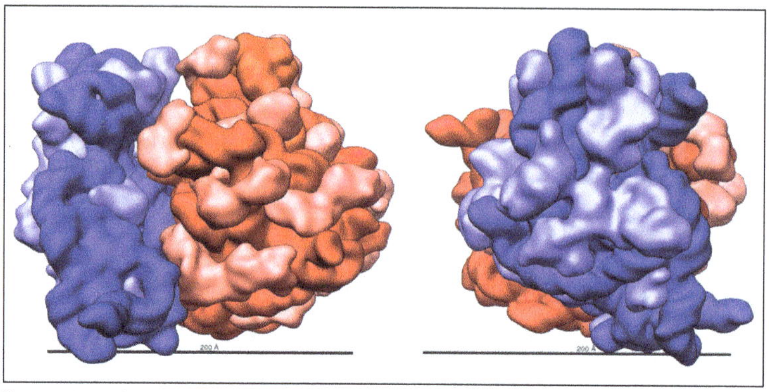

Three-dimensional views of the ribosome, showing rRNA in dark blue (small subunit) and dark red (large subunit). Lighter colors represent ribosomal proteins.

Ribosomal ribonucleic acid (rRNA) is the RNA component of the ribosome, which is essential for protein synthesis in all living organisms. rRNA is the predominant RNA in most cells, composing around 80% of cellular RNA. Ribosomes are approximately 60% rRNA and 40% protein by weight. A ribosome contains two subunits, the large ribosomal subunit (LSU) and small ribosomal subunit (SSU).

Prokaryotic ribosomes contain three rRNAs, which are the 23S and 5S rRNAs in the LSU and the 16S rRNA in the SSU. The prokaryotic ribosome contains around 50 ribosomal proteins.

Eukaryotic ribosomes and rRNAs are larger and more polymorphic than those of prokaryotes. In yeast, the LSU contains the 5S, 5.8S and 28S rRNAs. The combined 5.8S and 28S are roughly equivalent to the prokaryotic 23S rRNA, except for expansion segments (ESs) that are localized to the surface of the ribosome. The SSU contains the 18S rRNA, which also contains ESs. SSU ESs are generally smaller than LSU ESs.

The LSU rRNA has been called a ribozyme, because ribosomal protein does not penetrate into the catalytic site of the ribosome (the peptidyl transferase center, PTC). However, rRNA has not been shown to be catalytic in the absence of proteins. The SSU rRNA decodes the mRNA in the decoding center (DC). Ribosomal proteins do not penetrate into the DC.

SSU and LSU rRNA sequences are widely used for working out evolutionary relationships among organisms, since they are of ancient origin, are found in all known forms of life, and are resistant to horizontal gene transfer. The canonical tree of life is the lineage of the translation system.

Structure

During translation, tRNA is sandwiched between the small and large ribosomal subunits. In the SSU, the mRNA interacts with the anticodons of the tRNA. In the LSU, the amino acid acceptor stem of the tRNA interacts with the LSU rRNA. The ribosome catalyzes ester-amide exchange, transferring the C-terminus of a nascent peptide from a tRNA to the amino group of an amino acid.

A ribosome has three tRNA binding sites called the A, P, and E sites:

- The A site contains an aminoacyl-tRNA (a tRNA esterified to an amino acid on the 3' end).
- The P site contains a tRNA esterified to the nascent peptide. The free amino (NH_2) group of the A site tRNA attacks the ester linkage of P site tRNA, causing transfer of the nascent peptide to the amino acid in the A site. This reaction is takes place in the peptidyl transferase center.
- The E site contains a tRNA that has been discharged, with a free 3' end (with no amino acid or nascent peptide).

A single mRNA can be translated simultaneously by multiple ribosomes.

Subunits and Ribosomal RNA Genes

Both prokaryotic and eukaryotic ribosomes can be broken down into two subunits (the S in 16S represents Svedberg units), nt= length in nucleotides of the respective rRNAs, for exemplary species *Escherichia coli* (prokaryote) and human (eukaryote):

Type	Size	Large subunit (LSU rRNA)	Small subunit (SSU rRNA)
prokaryotic	70S	50S (5S : 120 nt, 23S : 2906 nt)	30S (16S : 1542 nt)
eukaryotic	80S	60S (5S : 121 nt, 5.8S : 156 nt, 28S : 5070 nt)	40S (18S : 1869 nt)

Note that the S units of the subunits (or the rRNAs) cannot simply be added because they represent measures of sedimentation rate rather than of mass. The sedimentation rate of each subunit is affected by its shape, as well as by its mass. The nt units can be added as these represent the integer number of units in the linear rRNA polymers (for example, the total length of the human rRNA = 7216 nt).

In Prokaryotes

In prokaryotes a small 30S ribosomal subunit contains the 16S ribosomal RNA. The large 50S ribosomal subunit contains two rRNA species (the 5S and 23S ribosomal RNAs).

Bacterial 16S ribosomal RNA, 23S ribosomal RNA, and 5S rRNA genes are typically organized as a co-transcribed operon. There is an internal transcribed spacer between 16S and 23S rRNA genes. There may be one or more copies of the operon dispersed in the genome (for example, *Escherichia coli* has seven).

Archaea contains either a single rDNA operon or multiple copies of the operon.

The 3' end of the 16S ribosomal RNA (in a ribosome) recognizes a sequence on the 5' end of mRNA called the Shine-Dalgarno sequence.

In Eukaryotes

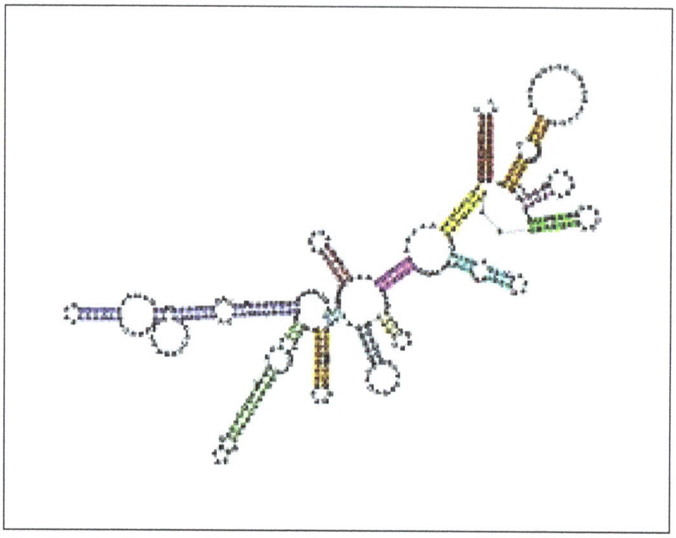

Small subunit ribosomal RNA, 5' domain taken from the Rfam database. This example is RF00177.

In contrast, eukaryotes generally have many copies of the rRNA genes organized in tandem repeats. In humans, approximately 300–400 repeats are present in five clusters, located on chromosomes 13 (RNR1), 14 (RNR2), 15 (RNR3), 21 (RNR4) and 22 (RNR5). Sequence variation in rRNA has been observed both within and between human individuals and certain variants are expressed in a tissue-specific manner in mice. Because of their special structure and transcription behaviour, rRNA gene clusters are commonly called "ribosomal DNA" (note that the term seems to imply that ribosomes contain DNA, which is not the case).

The 18S rRNA in most eukaryotes is in the small ribosomal subunit, and the large subunit contains three rRNA species (the 5S, 5.8S and 28S in mammals, 25S in plants, rRNAs).

Mammalian cells have 2 mitochondrial (12S and 16S) rRNA molecules and 4 types of cytoplasmic rRNA (the 28S, 5.8S, 18S, and 5S subunits). The 28S, 5.8S, and 18S rRNAs are encoded by a single transcription unit (45S) separated by 2 internally transcribed spacers. One of them corresponds to the one found in bacteria and archaea, and the other is an insertion into what was the 23S rRNA in prokaryotes. The 45S rDNA is organized into 5 clusters (each has 30–40 repeats) on chromosomes 13, 14, 15, 21, and 22. These are transcribed by RNA polymerase I. 5S occurs in tandem arrays (~200–300 true 5S genes and many dispersed pseudogenes), the largest one on the chromosome 1q41-42. 5S rRNA is transcribed by RNA polymerase III.

The tertiary structure of the small subunit ribosomal RNA (SSU rRNA) has been resolved by X-ray crystallography. The secondary structure of SSU rRNA contains 4 distinct domains—the 5', central, 3' major and 3' minor domains. A model of the secondary structure for the 5' domain (500-800 nucleotides) is shown.

Transcription

In Prokaryotes

In prokaryotic cells, each rRNA gene or operon is transcribed into a single RNA precursor that includes 16S, 23S, 5S rRNA and tRNA sequences along with transcribed spacers. The RNA processing then begins before the transcription is complete. During processing reactions, the rRNAs and tRNAs are released as separate molecules.

In Eukaryotes

The genes coding for 18S, 28S and 5.8S rRNA, located in the nucleolus organizer region, are transcribed into a large pre-rRNA molecule by RNA polymerase I. Each large pre-rRNA contains 18S, 28S and 5.8S sequences which are separated by external and internal transcribed spacer sequences. During processing reactions, the 18S, 28S and 5.8S rRNA are released as individual molecules. Processing reactions involve exo- and endo-nucleolytic cleavages guided by snoRNA (small nucleolar RNAs) in complex with proteins. The genes for 5S rRNA are located inside the nucleolus and are transcribed into pre-5S rRNA by RNA polymerase III. The pre-5S rRNA enters the nucleolus for processing and assembly with 28S and 5.8S rRNA to form the large subunit. 18S rRNA forms the small ribosomal subunits by combining with ribosomal proteins.

tmRNA

Transfer-messenger RNA (abbreviated tmRNA, also known as 10Sa RNA and by its genetic name SsrA) is a bacterial RNA molecule with dual tRNA-like and messenger RNA-like properties. The tmRNA forms a ribonucleoprotein complex (tmRNP) together with Small Protein B (SmpB), Elongation Factor Tu (EF-Tu), and ribosomal protein S1. In trans-translation, tmRNA and its associated proteins bind to bacterial ribosomes which have stalled in the middle of protein biosynthesis, for example when reaching the end of a messenger RNA which has lost its stop codon. The tmRNA is remarkably versatile: it recycles the stalled ribosome, adds a proteolysis-inducing tag to the unfinished polypeptide, and facilitates the degradation of the aberrant messenger RNA. In the majority of bacteria these functions are carried out by standard one-piece tmRNAs. In other bacterial species, a permuted ssrA gene produces a two-piece tmRNA in which two separate RNA chains are joined by base-pairing.

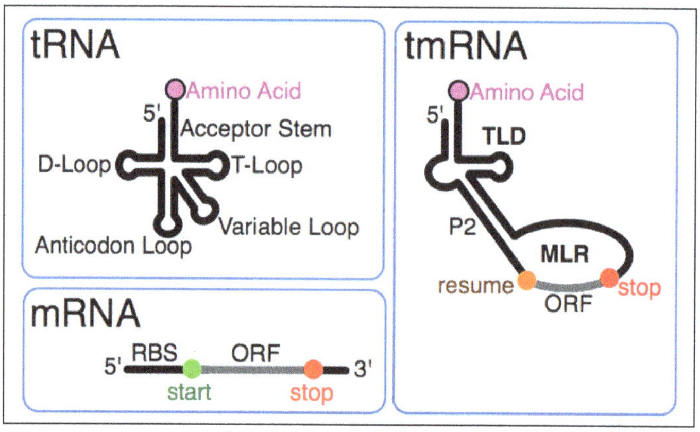

tmRNA combines features of tRNA and mRNA.

Discovery and Early Work

tmRNA was first designated 10Sa RNA after a mixed "10S" electrophoretic fraction of *Escherichia coli* RNA was further resolved into tmRNA and the similarly-sized RNase P RNA (10Sb). The presence of pseudouridine in the mixed 10S RNA hinted that tmRNA has modified bases found also in tRNA. The similarity at the 3' end of tmRNA to the T stem-loop of tRNA was first recognized upon sequencing *ssrA* from *Mycobacterium tuberculosis*. Subsequent sequence comparison revealed the full tRNA-like domain (TLD) formed by the 5' and 3' ends of tmRNA, including the acceptor stem with elements like those in alanine tRNA that promote its aminoacylation by alanine-tRNA ligase. It also revealed differences from tRNA: the anticodon arm is missing in tmRNA, and the D arm region is a loop without base pairs.

Structure

Secondary Structure of the Standard One-piece tmRNAs

Secondary structure of *E. coli* tmRNA. Shown are the 5' and 3' ends of the 363-nucleotide RNA chain numbered in increments of ten. Short lines indicate Watson-Crick pairings (G-C and A-U); dots are G-U pairings. Prominent are the tRNA-like domain (TLD), the messenger RNA-like region (MLR), and the four pseudoknots (pk1 to pk4). The MLR encodes the tag peptide between resume and stop codons. RNA helices (numbered one to 12) and their sections (letters) are gray.

The complete *E. coli* tmRNA secondary structure was elucidated by comparative sequence analysis and structural probing. Watson-Crick and G-U base pairs were identified by comparing the bacterial tmRNA sequences using automated computational methods in combination with manual alignment procedures. The accompanying figure shows the base pairing pattern of this prototypical tmRNA, which is organized into 12 phylogenetically supported helices (also called pairings P1 to P12), some divided into helical segments.

A prominent feature of every tmRNA is the conserved tRNA-like domain (TLD), composed of helices 1, 12, and 2a (analogs of the tRNA acceptor stem, T-stem and variable stem, respectively), and containing the 5' monophosphate and alanylatable 3' CCA ends. The mRNA-like region (MLR) is in standard tmRNA a large loop containing pseudoknots and a coding sequence (CDS) for the tag peptide, marked by the resume codon and the stop codon. The encoded tag peptide (ANDENYALAA in *E. coli*) varies among bacteria, perhaps depending on the set of proteases and adaptors available.

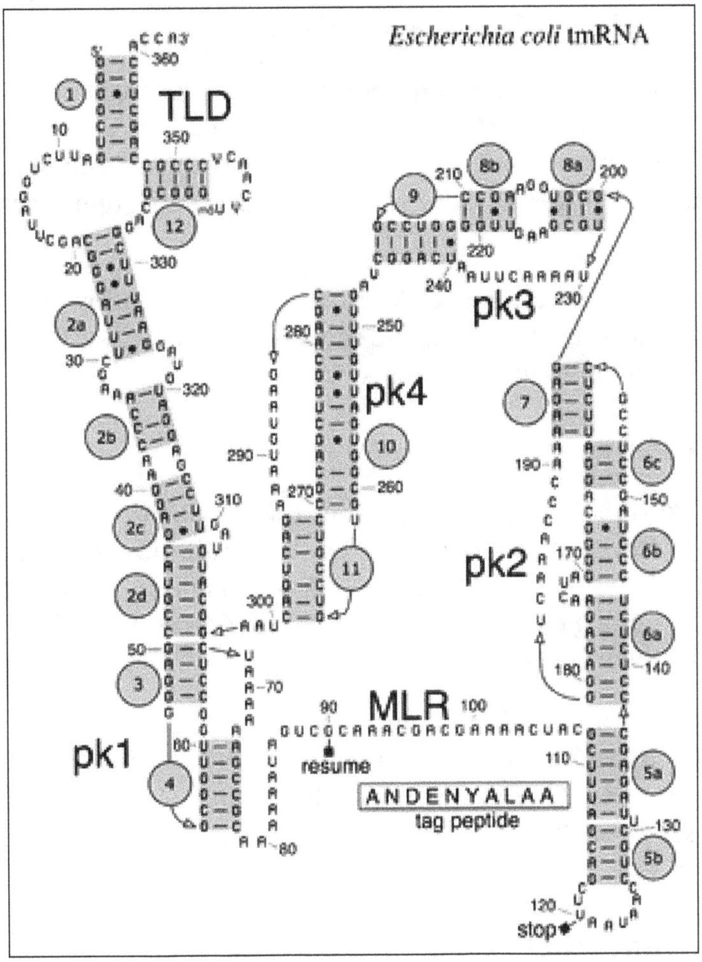

tmRNAs typically contain four pseudoknots, one (pk1) upstream of the tag peptide CDS, and the other three pseudoknots (pk2 to pk4) downstream of the CDS. The pseudoknot regions, although generally conserved, are evolutionarily plastic. For example, in the (one-piece) tmRNAs of cyanobacteria, pk4 is substituted with two tandemly arranged smaller pseudoknots. This suggests that tmRNA folding outside the TLD can be important, yet the pseudoknot region lacks conserved residues and pseudoknots are among the first structures to be lost as *ssrA* sequences diverge in plastid and endosymbiont lineages. Base pairing in the three-pseudoknot region of *E. coli* tmRNA is disrupted during *trans*-translation.

Two-piece TmRNAs

Circularly permuted *ssrA* has been reported in three major lineages: i) all alphaproteobacteria and the primitive mitochondria of jakobid protists, ii) two disjoint groups of cyanobacteria (*Gloeobacter* and a clade containing *Prochlorococcus* and many *Synechococcus*), and iii) some members of the betaproteobacteria (*Cupriavidus* and some Rhodocyclales). All produce the same overall two-piece (acceptor and coding pieces) form, equivalent to the standard form nicked downstream of the reading frame. None retain more than two pseudoknots compared to the four (or more) of standard tmRNA.

Alphaproteobacteria have two signature sequences: replacement of the typical T-loop sequence TΨCRANY with GGCRGUA, and the sequence AACAGAA in the large loop of the 3´-terminal pseudoknot. In mitochondria, the MLR has been lost, and a remarkable re-permutation of mitochondrial *ssrA* results in a small one-piece product in *Jakoba libera*.

The cyanobacteria provide the most plausible case for evolution of a permuted gene from a standard gene, due to remarkable sequence similarities between the two gene types as they occur in different *Synechococcus* strains.

tmRNA Processing

Most tmRNAs are transcribed as larger precursors which are processed much like tRNA. Cleavage at the 5´ end is by ribonuclease P. Multiple exonucleases can participate in the processing of the 3´ end of tmRNA, although RNase T and RNase PH are most effective. Depending on the bacterial species, the 3'-CCA is either encoded or added by tRNA nucleotidyltransferase.

Similar processing at internal sites of permuted precursor tmRNA explains its physical splitting into two pieces. The two-piece tmRNAs have two additional ends whose processing must be considered. For alphaproteobacteria, one 5´ end is the unprocessed start site of transcription. The far 3´ end may in some cases be the result of rho-independent termination.

Three-dimensional Structures

Cartoon ribbon structure of the tRNA-like domain of tmRNA. The domain consists of the 3' and 5' ends of the tmRNA.

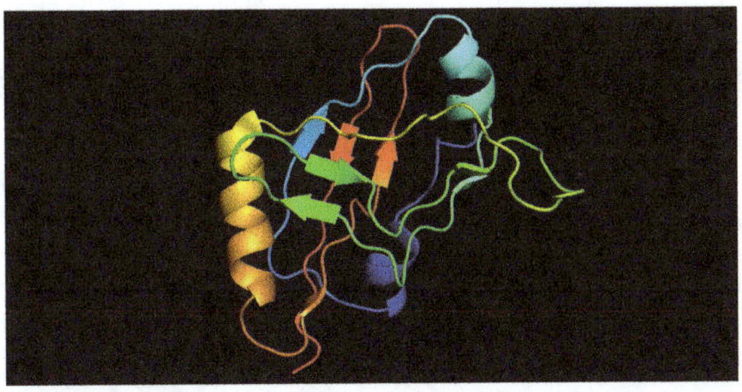

Cartoon ribbon structure of the tmRNA dedicated binding protein, SmpB.

High-resolution structures of the complete tmRNA molecules are currently unavailable and may be difficult to obtain due the inherent flexibility of the MLR. In 2007, the crystal structure of the *Thermus thermophilus* TLD bound to the SmpB protein was obtained at 3 Å resolution. This structure shows that SmpB mimics the D stem and the anticodon of a canonical tRNA whereas helical section 2a of tmRNA corresponds to the variable arm of tRNA. A cryo-electron microscopy study of tmRNA at an early stage of *trans*-translation shows the spatial relationship between the ribosome and the tmRNP (tmRNA bound to the EF-Tu protein). The TLD is located near the GTPase-associated center in the 50S ribosomal subunit; helix 5 and pseudoknots pk2 to pk4 form an arc around the beak of the 30S ribosomal subunit.

Trans-translation

Coding by tmRNA was discovered in 1995 when Simpson and coworkers overexpressed the mouse cytokine IL-6 in *E. coli* and found multiple truncated cytokine-derived peptides each tagged at the carboxyl termini with the same 11-amino acid residue extension (A)ANDENYALAA. With the exception of the N-terminal alanine, which comes from the 3' end of tmRNA itself, this tag sequence was traced to a short open reading frame in *E. coli* tmRNA. Recognizing that the tag peptide confers proteolysis, the *trans*-translation model for tmRNA action was proposed.

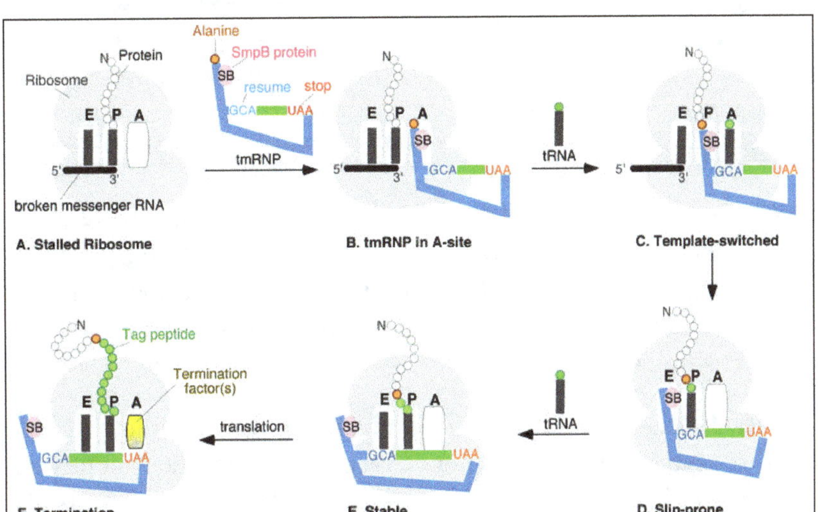

Trans-translation stages A through F. A ribosome with its RNA binding sites, designated E, P, and A, is stuck near the 3' end of a broken mRNA. The tmRNP binds to the A-site, allowing the ribosome to switch templates from the broken message onto the open reading frame of the tmRNA via the resume codon (blue GCA). Regular translation eventually resumes. Upon reaching the tmRNA stop codon (red UAA), a hybrid protein with a proteolysis tag (green beads) is released.

While details of the *trans*-translation mechanism are under investigation it is generally agreed that tmRNA first occupies the empty A site of the stalled ribosome. Subsequently, the ribosome moves from the 3' end of the truncated messenger RNA onto the resume codon of the MLR, followed by a slippage-prone stage from where translation continues normally until the in-frame tmRNA stop codon is encountered. *Trans-translation* is essential in some bacterial species, whereas other bacteria require tmRNA to survive when subjected to stressful growth conditions. Depending on the organism, the tag peptide may be recognized by a variety of proteases or protease adapters.

Mobile Genetic Elements and the tmRNA Gene

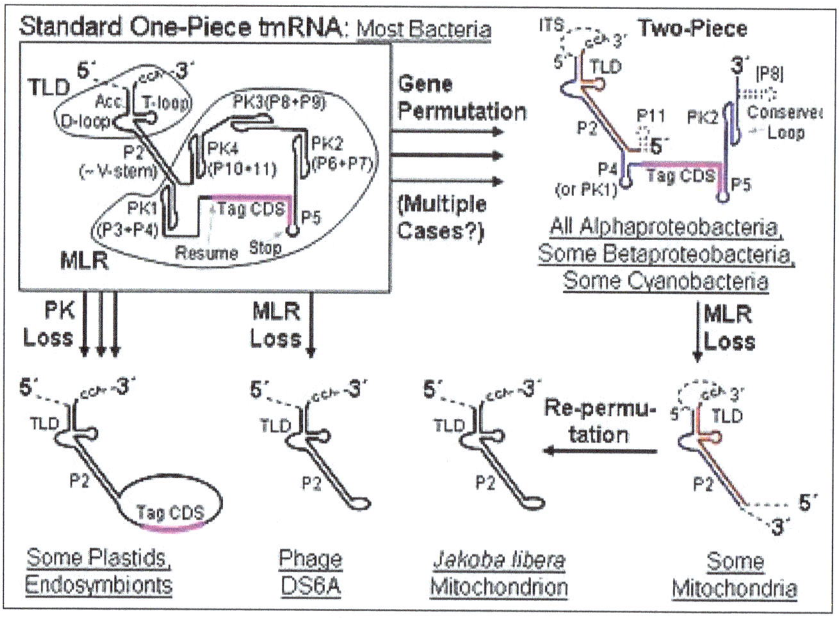

Precursor RNAs are shown, whose dashed portions are excised during maturation. The permuted genes produce both an acceptor piece (red) and coding piece (blue); dotted lines mark secondary structures not always present. Abbreviations: TLD, tRNA-like domain; MLR, mRNA-like region; ITS, internal transcribed spacer; P, paired region; PK, pseudoknot; RF, reading frame.

ssrA is both a target for some mobile DNAs and a passenger on others. It has been found interrupted by three types of mobile elements. By different strategies none of these disrupt gene function: Group I introns remove themselves by self-splicing, rickettsial palindromic elements (RPEs) insert in innocuous sites, and integrase-encoding genomic islands split their target *ssrA* yet restore the split-off portion.

Non-chromosomal *ssrA* was first detected in a genomic survey of mycobacteriophages (in 10% of the phages). Other mobile elements including plasmids and genomic islands have been found bearing *ssrA*. One interesting case is *Rhodobacter sphaeroides* ATCC 17025, whose native tmRNA gene is disrupted by a genomic island; unlike all other genomic islands in tmRNA (or tRNA) genes this island has inactivated the native target gene without restoration, yet compensates by carrying its own tmRNA gene. A very unusual relative of *ssrA* is found in the lytic mycobacteriophage DS6A, that encodes little more than the TLD.

Mitochondrial tmRNAs (ssrA Gene)

A mitochondrion-encoded, structurally reduced form of tmRNA (mt-tmRNA) was first postulated for the jakobid flagellate *Reclinomonas americana*. Subsequently, the presence of a mitochondrial gene (*ssrA*) coding for tmRNA, as well as transcription and RNA processing sites were confirmed for all but one member of jakobids. Functional evidence, i.e., mt-tmRNA Aminoacylation with alanine, is available for *Jakoba libera*. More recently, *ssrA* was also identified in mitochondrial genomes of oomycetes. Like in α-Proteobacteria (the ancestors of mitochondria), mt-tmRNAs are circularly permuted, two-piece RNA molecules, except in *Jakoba libera* where the gene has reverted to encoding a one-piece tmRNA conformation.

Identification of ssrA in Mitochondrial Genomes

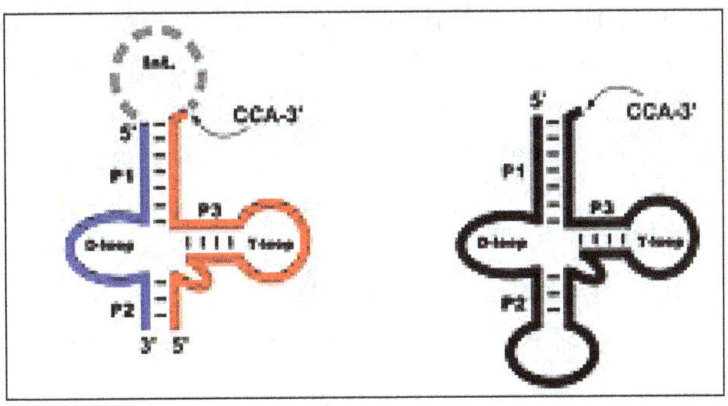

Secondary structure models for mt-tmRNAs. (A) The two-piece tmRNA in oomycetes and jakobids except J. libera, After removing the intervening sequence (Int.; dashed ark) by RNA processing enzymes, the two resulting RNA pieces (blue and red lines) remain together through base pairing. (B) The standard one-piece tmRNA in J. libera that most likely emerged secondarily through gene rearrangement. The three pairing regions (P1, P2 and P3) and the position of post-transcriptionally added 3' CCA are indicated.

Mitochondrial tmRNA genes were initially recognized as short sequences that are conserved among jakobids and that have the potential to fold into a distinct tRNA-like secondary structure. With the availability of nine complete jakobid mtDNA sequences, and a significantly improved covariance search tool (Infernal;), a covariance model has been developed based on jakobid mitochondrial tmRNAs, which identified mitochondrial *ssrA* genes also in oomycete. At present, a total of 34 oomycete mt-tmRNAs have been detected across six genera: Albugo, Bremia, Phytophthora, Pseudoperonospora, Pythium and Saprolegnia. A covariance model built with both jakobid and oomycete sequences is now available at Rfam under the name 'mt-tmRNA'.

mt-tmRNA Structure

The standard bacterial tmRNA consists of a tRNA(Ala)-like domain (allowing addition of a non-encoded alanine to mRNAs that happen to lack a stop coding), and an mRNA-like domain coding for a protein tag that destines the polypeptide for proteolysis. The mRNA-like domain was lost in mt-tmRNAs. Comparative sequence analysis indicates features typical for mt-tmRNAs. Most conserved is the primary sequence of the amino acyl acceptor stem. This portion of the molecule has an invariable A residue in the discriminator position and a G-U pair at position 3 (except in *Seculamonas ecuadoriensis*, which has a G-C pair); this position is the recognition site for alanyl tRNA synthase. P2 is a helix of variable length (3 to 10 base pairs) and corresponds to the anticodon stem of tRNAs, yet without an anticodon loop (as not required for tmRNA function). P2 stabilizes the tRNA-like structure, but four nucleotides invariant across oomycetes and jakobids suggest an additional, currently unidentified function. P3 has five base pairs and corresponds to the T-arm of tRNAs, yet with different consensus nucleotides both in the paired region and the loop. The T-loop sequence is conserved across oomycetes and jakobid, with only few deviations (e.g., *Saprolegnia ferax*). Finally, instead of the tRNA-like D-stem with a shortened three-nucleotide D-loop characteristic for bacterial tmRNAs, mitochondrial counterparts have a highly variable 5 to 14-nt long loop. The intervening sequence (Int.) of two-piece mt-tmRNAs is A+U rich and of irregular length (4-34 nt)). For secondary structure models of one- and two-piece mt-tmRNAs.

mt-tmRNA Processing and Expression

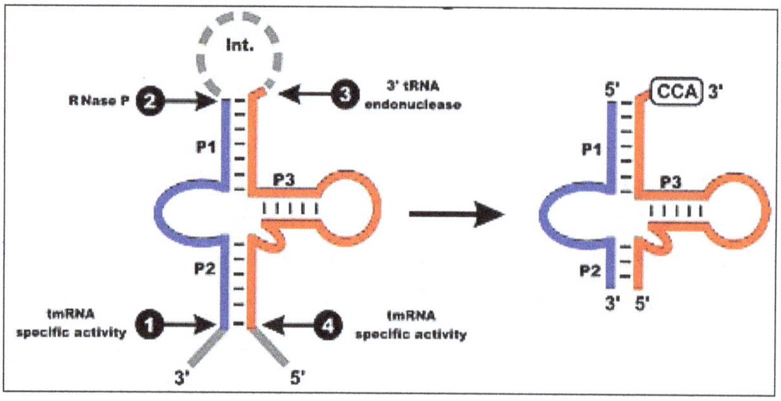

Processing of two-piece mt-tmRNA. The four major RNA processing sites are numbered (1-4). Processing at sites 1 and 4 is thought to occur by a tmRNA-specific activity, site 2 by RNase P and site 3 by a 3' tRNA endonuclease processing. Nucleotides cleaved from the precursor are in gray; the post-transcriptionally added CCA is boxed.

RNA-Seq data of *Phytophthora sojae* show an expression level similar to that of neighboring mitochondrial tRNAs, and four major processing sites confirm the predicted termini of mature mt-tmRNA. The tmRNA precursor molecule is likely processed by RNase P and a tRNA 3' processing endonuclease; the latter activity is assumed to lead to the removal of the intervening sequence. Following the addition of CCA at the 3' discriminator nucleotide, the tmRNA can be charged by alanyl-tRNA synthetase with alanine.

Polyadenylation

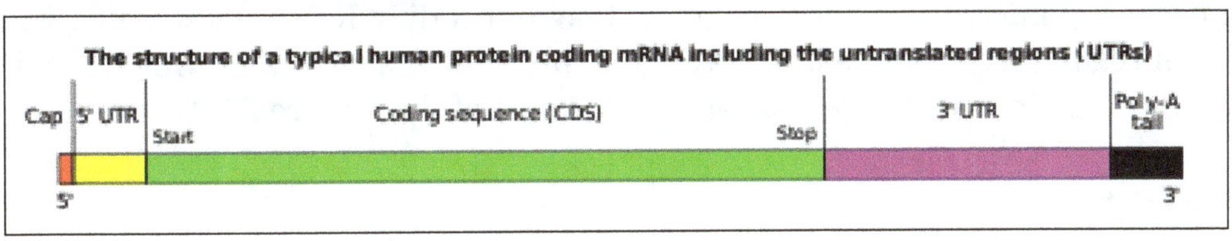

Typical structure of a mature eukaryotic mRNA.

Polyadenylation is the addition of a poly(A) tail to a messenger RNA. The poly(A) tail consists of multiple adenosine monophosphates; in other words, it is a stretch of RNA that has only adenine bases. In eukaryotes, polyadenylation is part of the process that produces mature messenger RNA (mRNA) for translation. It, therefore, forms part of the larger process of gene expression.

The process of polyadenylation begins as the transcription of a gene terminates. The 3'-most segment of the newly made pre-mRNA is first cleaved off by a set of proteins; these proteins then synthesize the poly(A) tail at the RNA's 3' end. In some genes these proteins add a poly(A) tail at one of several possible sites. Therefore, polyadenylation can produce more than one transcript from a single gene (alternative polyadenylation), similar to alternative splicing.

The poly(A) tail is important for the nuclear export, translation, and stability of mRNA. The tail is shortened over time, and, when it is short enough, the mRNA is enzymatically degraded. However, in a few cell types, mRNAs with short poly(A) tails are stored for later activation by re-polyadenylation in the cytosol. In contrast, when polyadenylation occurs in bacteria, it promotes RNA degradation. This is also sometimes the case for eukaryotic non-coding RNAs.

mRNA molecules in both prokaryotes and eukaryotes have polyadenylated 3'-ends, with the prokaryotic poly(A) tails generally shorter and less mRNA molecules polyadenylated.

Nuclear Polyadenylation

Function

In nuclear polyadenylation, a poly(A) tail is added to an RNA at the end of transcription. On mRNAs, the poly(A) tail protects the mRNA molecule from enzymatic degradation in the cytoplasm and aids in transcription termination, export of the mRNA from the nucleus, and translation. Almost all eukaryotic mRNAs are polyadenylated, with the exception of animal replication-dependent histone mRNAs. These are the only mRNAs in eukaryotes that lack a poly(A) tail, ending instead in a stem-loop structure followed by a purine-rich sequence, termed histone downstream element, that directs where the RNA is cut so that the 3' end of the histone mRNA is formed.

Many eukaryotic non-coding RNAs are always polyadenylated at the end of transcription. There are small RNAs where the poly(A) tail is seen only in intermediary forms and not in the mature RNA as the ends are removed during processing, the notable ones being microRNAs. But, for many long noncoding RNAs – a seemingly large group of regulatory RNAs that, for example, includes the RNA Xist, which mediates X chromosome inactivation – a poly(A) tail is part of the mature RNA.

Mechanism

The processive polyadenylation complex in the nucleus of eukaryotes works on products of RNA polymerase II, such as precursor mRNA. Here, a multi-protein complex cleaves the 3'-most part of a newly produced RNA and polyadenylates the end produced by this cleavage. The cleavage is catalysed by the enzyme CPSF and occurs 10–30 nucleotides downstream of its binding site. This site often has the polyadenylation signal sequence AAUAAA on the RNA, but variants of it that bind more weakly to CPSF exist. Two other proteins add specificity to the binding to an RNA: CstF and CFI. CstF binds to a GU-rich region further downstream of CPSF's site. CFI recognises a third site on the RNA (a set of UGUAA sequences in mammals) and can recruit CPSF even if the AAUAAA sequence is missing. The polyadenylation signal – the sequence motif recognised by the RNA cleavage complex – varies between groups of eukaryotes. Most human polyadenylation sites contain the AAUAAA sequence, but this sequence is less common in plants and fungi.

The RNA is typically cleaved before transcription termination, as CstF also binds to RNA polymerase II. Through a poorly understood mechanism, it signals for RNA polymerase II to slip off of the transcript. Cleavage also involves the protein CFII, though it is unknown how. The cleavage site associated with a polyadenylation signal can vary up to some 50 nucleotides.

When the RNA is cleaved, polyadenylation starts, catalysed by polyadenylate polymerase. Polyadenylate polymerase builds the poly(A) tail by adding adenosine monophosphate units from adenosine triphosphate to the RNA, cleaving off pyrophosphate.

Another protein, PAB2, binds to the new, short poly(A) tail and increases the affinity of polyadenylate polymerase for the RNA. When the poly(A) tail is approximately 250 nucleotides long the enzyme can no longer bind to CPSF and polyadenylation stops, thus determining the length of the poly(A) tail. CPSF is in contact with RNA polymerase II, allowing it to signal the polymerase to terminate transcription. When RNA polymerase II reaches a "termination sequence" ($^{5'}$TTTATT$^{3'}$ on the DNA template and $^{5'}$AAUAAA$^{3'}$) on the primary transcript), the end of transcription is signaled. The polyadenylation machinery is also physically linked to the spliceosome, a complex that removes introns from RNAs.

Downstream Effects

The poly(A) tail acts as the binding site for poly(A)-binding protein. Poly(A)-binding protein promotes export from the nucleus and translation, and inhibits degradation. This protein binds to the poly(A) tail prior to mRNA export from the nucleus and in yeast also recruits poly(A) nuclease, an enzyme that shortens the poly(A) tail and allows the export of the mRNA. Poly(A)-binding protein is exported to the cytoplasm with the RNA. mRNAs that are not exported are degraded by the exosome. Poly(A)-binding protein also can bind to, and thus recruit, several proteins that affect translation, one of these is initiation factor-4G, which in turn recruits the 40S ribosomal subunit. However, a poly(A) tail is not required for the translation of all mRNAs. Further, poly(A) tailing (oligo-adenylation) can determine the fate of RNA molecules that are usually not poly(A)-tailed (such as (small) non-coding (sn)RNAs etc.) and thereby induce their RNA decay.

Deadenylation

In eukaryotic somatic cells, the poly(A) tail of most mRNAs in the cytoplasm gradually get shorter, and mRNAs with shorter poly(A) tail are translated less and degraded sooner. However, it can take many hours before an mRNA is degraded. This deadenylation and degradation process can be accelerated by microRNAs complementary to the 3' untranslated region of an mRNA. In immature egg cells, mRNAs with shortened poly(A) tails are not degraded, but are instead stored without being translated. They are then activated by cytoplasmic polyadenylation after fertilisation, during egg activation.

In animals, poly(A) ribonuclease (PARN) can bind to the 5' cap and remove nucleotides from the poly(A) tail. The level of access to the 5' cap and poly(A) tail is important in controlling how soon the mRNA is degraded. PARN deadenylates less if the RNA is bound by the initiation factors 4E (at the 5' cap) and 4G (at the poly(A) tail), which is why translation reduces deadenylation. The rate of deadenylation may also be regulated by RNA-binding proteins. Once the poly(A) tail is removed, the decapping complex removes the 5' cap, leading to a degradation of the RNA. Several other enzymes that seem to be involved in deadenylation have been identified in yeast.

Alternative Polyadenylation

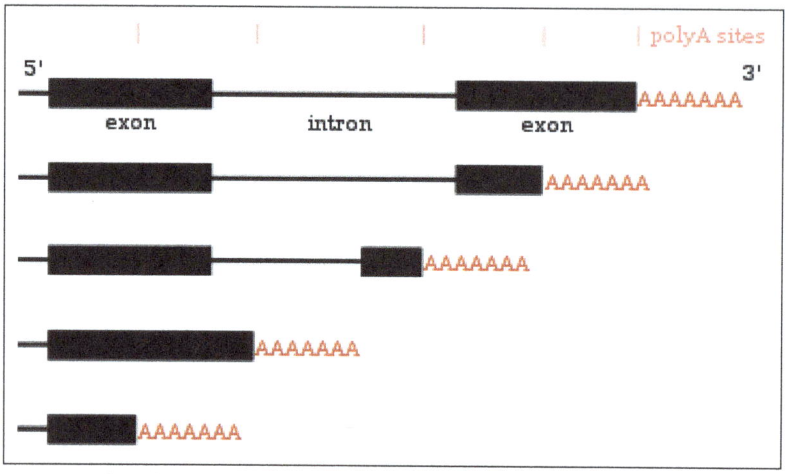

Results of using different polyadenylation sites on the same gene.

Many protein-coding genes have more than one polyadenylation site, so a gene can code for several mRNAs that differ in their 3' end. Since alternative polyadenylation changes the length of the 3' untranslated region, it can change which binding sites for microRNAs the 3' untranslated region contains. MicroRNAs tend to repress translation and promote degradation of the mRNAs they bind to, although there are examples of microRNAs that stabilise transcripts. Alternative polyadenylation can also shorten the coding region, thus making the mRNA code for a different protein, but this is much less common than just shortening the 3' untranslated region.

The choice of poly(A) site can be influenced by extracellular stimuli and depends on the expression of the proteins that take part in polyadenylation. For example, the expression of CstF-64, a subunit of cleavage stimulatory factor (CstF), increases in macrophages in response to lipopolysaccharides (a group of bacterial compounds that trigger an immune response). This results in the selection of weak poly(A) sites and thus shorter transcripts. This removes regulatory elements in the 3' untranslated regions of mRNAs for defense-related products like lysozyme and TNF-α. These mRNAs then have longer half-lives and produce more of these proteins. RNA-binding proteins other than those in the polyadenylation machinery can also affect whether a polyadenyation site is used, as can DNA methylation near the polyadenylation signal.

Cytoplasmic Polyadenylation

There is polyadenylation in the cytosol of some animal cell types, namely in the germ line, during early embryogenesis and in post-synaptic sites of nerve cells. This lengthens the poly(A) tail of an mRNA with a shortened poly(A) tail, so that the mRNA will be translated. These shortened poly(A) tails are often less than 20 nucleotides, and are lengthened to around 80–150 nucleotides.

In the early mouse embryo, cytoplasmic polyadenylation of maternal RNAs from the egg cell allows the cell to survive and grow even though transcription does not start until the middle of the 2-cell stage (4-cell stage in human). In the brain, cytoplasmic polyadenylation is active during learning and could play a role in long-term potentiation, which is the strengthening of the signal transmission from a nerve cell to another in response to nerve impulses and is important for learning and memory formation.

Cytoplasmic polyadenylation requires the RNA-binding proteins CPSF and CPEB, and can involve other RNA-binding proteins like Pumilio. Depending on the cell type, the polymerase can be the same type of polyadenylate polymerase (PAP) that is used in the nuclear process, or the cytoplasmic polymerase GLD-2.

Tagging for Degradation in Eukaryotes

For many non-coding RNAs, including tRNA, rRNA, snRNA, and snoRNA, polyadenylation is a way of marking the RNA for degradation, at least in yeast. This polyadenylation is done in the nucleus by the TRAMP complex, which maintains a tail that is around 4 nucleotides long to the 3' end. The RNA is then degraded by the exosome. Poly(A) tails have also been found on human rRNA fragments, both the form of homopolymeric (A only) and heterpolymeric (mostly A) tails.

In Prokaryotes and Organelles

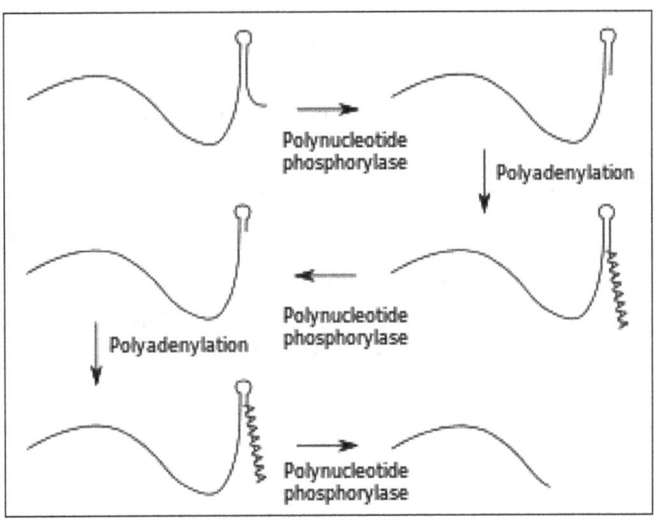

Polyadenylation in bacteria helps polynucleotide phosphorylase degrade past secondary structure.

In many bacteria, both mRNAs and non-coding RNAs can be polyadenylated. This poly(A) tail promotes degradation by the degradosome, which contains two RNA-degrading enzymes: polynucleotide phosphorylase and RNase E. Polynucleotide phosphorylase binds to the 3' end of RNAs and the 3' extension provided by the poly(A) tail allows it to bind to the RNAs whose secondary structure would otherwise block the 3' end. Successive rounds of polyadenylation and degradation of the 3' end by polynucleotide phosphorylase allows the degradosome to overcome these secondary structures. The poly(A) tail can also recruit RNases that cut the RNA in two. These bacterial poly(A) tails are about 30 nucleotides long.

In as different groups as animals and trypanosomes, the mitochondria contain both stabilising and destabilising poly(A) tails. Destabilising polyadenylation targets both mRNA and noncoding RNAs. The poly(A) tails are 43 nucleotides long on average. The stabilising ones start at the stop codon, and without them the stop codon (UAA) is not complete as the genome only encodes the U or UA part. Plant mitochondria have only destabilising polyadenylation, and yeast mitochondria have no polyadenylation at all.

While many bacteria and mitochondria have polyadenylate polymerases, they also have another type of polyadenylation, performed by polynucleotide phosphorylase itself. This enzyme is found in bacteria, mitochondria, plastids and as a constituent of the archaeal exosome (in those archaea that have an exosome). It can synthesise a 3' extension where the vast majority of the bases are adenines. Like in bacteria, polyadenylation by polynucleotide phosphorylase promotes degradation of the RNA in plastids and likely also archaea.

Evolution

Although polyadenylation is seen in almost all organisms, it is not universal. However, the wide distribution of this modification and the fact that it is present in organisms from all three domains of life implies that the last universal common ancestor of all living organisms, it is presumed, had some form of polyadenylation system. A few organisms do not polyadenylate mRNA, which implies that they have lost their polyadenylation machineries during evolution. Although no examples of eukaryotes that lack polyadenylation are known, mRNAs from the bacterium Mycoplasma gallisepticum and the salt-tolerant archaean Haloferax volcanii lack this modification.

The most ancient polyadenylating enzyme is polynucleotide phosphorylase. This enzyme is part of both the bacterial degradosome and the archaeal exosome, two closely related complexes that recycle RNA into nucleotides. This enzyme degrades RNA by attacking the bond between the 3'-most nucleotides with a phosphate, breaking off a diphosphate nucleotide. This reaction is reversible, and so the enzyme can also extend RNA with more nucleotides. The heteropolymeric tail added by polynucleotide phosphorylase is very rich in adenine. The choice of adenine is most likely the result of higher ADP concentrations than other nucleotides as a result of using ATP as an energy currency, making it more likely to be incorporated in this tail in early lifeforms. It has been suggested that the involvement of adenine-rich tails in RNA degradation prompted the later evolution of polyadenylate polymerases (the enzymes that produce poly(A) tails with no other nucleotides in them).

Polyadenylate polymerases are not as ancient. They have separately evolved in both bacteria and eukaryotes from CCA-adding enzyme, which is the enzyme that completes the 3' ends of tRNAs. Its catalytic domain is homologous to that of other polymerases. It is presumed that the horizontal transfer of bacterial CCA-adding enzyme to eukaryotes allowed the archaeal-like CCA-adding enzyme to switch function to a poly(A) polymerase. Some lineages, like archaea and cyanobacteria, never evolved a polyadenylate polymerase.

Transcriptome

The human genome is made up of DNA (deoxyribonucleic acid), a long, winding molecule that contains the instructions needed to build and maintain cells. These instructions are spelled out in the form of "base pairs" of four different chemicals, organized into 20,000 to 25,000 genes. For the instructions to be carried out, DNA must be "read" and transcribed - in other words, copied - into RNA (ribonucleic acid). These gene readouts are called transcripts, and a transcriptome is a collection of all the gene readouts present in a cell.

There are various kinds of RNA. The major type, called messenger RNA (mRNA), plays a vital role in making proteins. In this process: mRNA is transcribed from genes; then the mRNA transcripts are delivered to ribosomes, the molecular machines located in the cell's cytoplasm; then the ribosomes read, or "translate," the sequence of chemical letters in the mRNA and assemble building blocks called amino acids into proteins.

DNA can also be transcribed into other types of RNA that do not code for proteins. Such transcripts may serve to influence cell structure and to regulate genes.

Importance of Transcriptome

An RNA sequence mirrors the sequence of the DNA from which it was transcribed. Consequently, by analyzing the entire collection of RNA sequences in a cell (the transcriptome) researchers can determine when and where each gene is turned on or off in the cells and tissues of an organism.

Depending on the technique used, it is often possible to count the number of transcripts to determine the amount of gene activity - also called gene expression - in a certain cell or tissue type.

In humans and other organisms, nearly every cell contains the same genes, but different cells show different patterns of gene expression. These differences are responsible for the many different properties and behaviors of various cells and tissues, both in health and disease.

By collecting and comparing transcriptomes of different types of cells, researchers can gain a deeper understanding of what constitutes a specific cell type, how that type of cell normally functions, and how changes in the normal level of gene activity may reflect or contribute to disease. In addition, transcriptomes may enable researchers to generate a comprehensive, genome-wide picture of what genes are active in which cells.

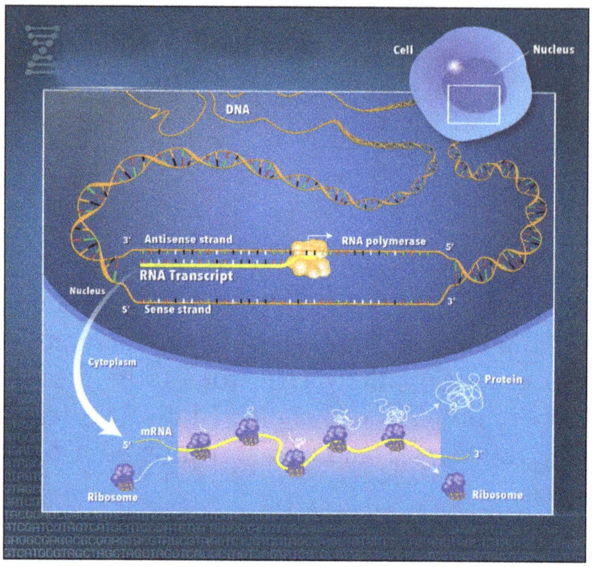

Use of Trancriptome

The function of most genes is not yet known. A search of a transcriptome database can give researchers a list of all the tissues in which a gene is expressed, providing clues about its possible function.

For example, if the transcriptome database shows that an unknown gene's expression levels are dramatically higher in cancer cells than in healthy cells, the unknown gene may play a role in cell growth. Or if an unknown gene is expressed in fat tissue but not in bone or muscle tissue, the unknown gene may be involved in fat storage or metabolism. In both instances, the transcriptome data give researchers a good place to start searching for a newly found gene's function.

The National Human Genome Research Institute (NHGRI), which is part of the National Institutes of Health (NIH), has participated in two projects that created transcriptome resources for researchers around the world-the Mammalian Gene Collection initiative and the Mouse Transcriptome Project.

The Mammalian Gene Collection initiative built a free, public library of human, mouse, and rat mRNA sequences. The project was led by NHGRI and the National Cancer Institute (NCI), also part of NIH. The mouse and rat are important models with which to study human biology.

The Mouse Transcriptome Project was an NIH-supported initiative that generated a free, public database of gene transcripts for many mouse tissues. These tissue-specific gene expression data, which are mapped to the mouse genome, are available in a searchable format in the Mouse Reference Transcriptome Database.

Several other transcriptome resources exist, including those in NIH programs such as the Genotype-Tissue Expression Project (GTEx) and the Encyclopedia of DNA Elements (ENCODE). GTEx is creating a catalog of human gene expression in a variety of different tissues. ENCODE researchers aim to characterize and understand the working parts of the genome, including the transcriptome. Both Novartis and the European Molecular Biology Laboratory have well-established gene expression databases.

Untranslated Regions of RNA

Untranslated regions (UTRs) are sections of the mRNA before the start codon and after the stop codon that are not translated, termed the five prime untranslated region (5' UTR) and three prime untranslated region (3' UTR), respectively. These regions are transcribed with the coding region and thus are exonic as they are present in the mature mRNA. Several roles in gene expression have been attributed to the untranslated regions, including mRNA stability, mRNA localization, and translational efficiency. The ability of a UTR to perform these functions depends on the sequence of the UTR and can differ between mRNAs. Genetic variants in 3' UTR have also been implicated in disease susceptibility because of the change in RNA structure and protein translation.

The stability of mRNAs may be controlled by the 5' UTR and 3' UTR due to varying affinity for RNA degrading enzymes called ribonucleases and for ancillary proteins that can promote or inhibit RNA degradation.

Translational efficiency, including sometimes the complete inhibition of translation, can be controlled by UTRs. Proteins that bind to either the 3' or 5' UTR may affect translation by influencing the ribosome's ability to bind to the mRNA. MicroRNAs bound to the 3' UTR also may affect translational efficiency or mRNA stability.

Cytoplasmic localization of mRNA is thought to be a function of the 3' UTR. Proteins that are needed in a particular region of the cell can also be translated there; in such a case, the 3' UTR may contain sequences that allow the transcript to be localized to this region for translation.

Some of the elements contained in untranslated regions form a characteristic secondary structure when transcribed into RNA. These structural mRNA elements are involved in regulating the mRNA. Some, such as the SECIS element, are targets for proteins to bind. One class of mRNA element, the riboswitches, directly bind small molecules, changing their fold to modify levels of transcription or translation. In these cases, the mRNA regulates itself.

Five Prime Untranslated Region

The 5′ untranslated region (5′ UTR) (also known as a leader sequence or leader RNA) is the region of an mRNA that is directly upstream from the initiation codon. This region is important for the regulation of translation of a transcript by differing mechanisms in viruses, prokaryotes and eukaryotes. While called untranslated, the 5′ UTR or a portion of it is sometimes translated into a protein product. This product can then regulate the translation of the main coding sequence of the mRNA. In many organisms, however, the 5′ UTR is completely untranslated, instead forming complex secondary structure to regulate translation. The 5′ UTR has been found to interact with proteins relating to metabolism; and proteins translate sequences within the 5′ UTR. In addition, this region has been involved in transcription regulation, such as the sex-lethal gene in Drosophila. Regulatory elements within 5′ UTRs have also been linked to mRNA export.

General Structure

Length

The 5′ UTR begins at the transcription start site and ends one nucleotide (nt) before the initiation sequence (usually AUG) of the coding region. In prokaryotes, the length of the 5′ UTR tends to be 3-10 nucleotides long, while in eukaryotes it tends to be anywhere from 100 to several thousand nucleotides long. For example, the *ste11* transcript in *Schizosaccharomyces pombe* has a 2273 nucleotide 5′ UTR while the *lac* operon in *Escherichia coli* only has 7 nucleotides in its 5′ UTR. The differing sizes are likely due to the complexity of the eukaryotic regulation which the 5′ UTR holds, as well as the larger preinitiation complex which must form to begin translation.

Elements

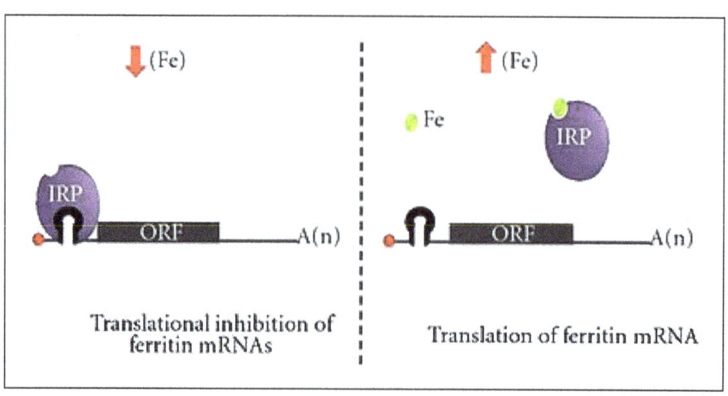

The binding of an IRP (Iron Regulatory Protein) to and IRE (Iron Response Element), which are hairpin loops, regulation translation.

The elements of a eukaryotic and prokaryotic 5′ UTR differ greatly. The prokaryotic 5′ UTR contains a ribosome binding site (RBS), also known as the Shine Dalgarno sequence (AGGAGGU), which is usually 3-10 base pairs upstream from the initiation codon. In contrast, the eukaryotic 5′ UTR contains the Kozak consensus sequence (ACCAUGG), which contains the initiation codon. The eukaryotic 5′ UTR also contains *cis*-acting regulatory elements called upstream open reading frames (uORFs) and upstream AUGs (uAUGs) and termination codons, which have a great impact on the regulation of translation. Unlike prokaryotes, 5′ UTRs can harbor introns in eukaryotes. In humans, ~35% of all genes harbor introns within the 5′ UTR.

Secondary Structure

As the 5′ UTR has a high GC content, secondary structures often occur within it. Hairpin loops are one such secondary structure that can be located within the 5′ UTR. These secondary structures also impact the regulation of translation.

Role in Translational Regulation

Prokaryotes

In prokaryotes, the initiation of translation occurs when IF-3, along with the 30S ribosomal subunit, bind to the Shine-Dalgarno sequence of the 5′ UTR. This then recruits many other proteins, such as the 50S ribosomal subunit, which allows for translation to begin. Each of these steps regulates the initiation of translation.

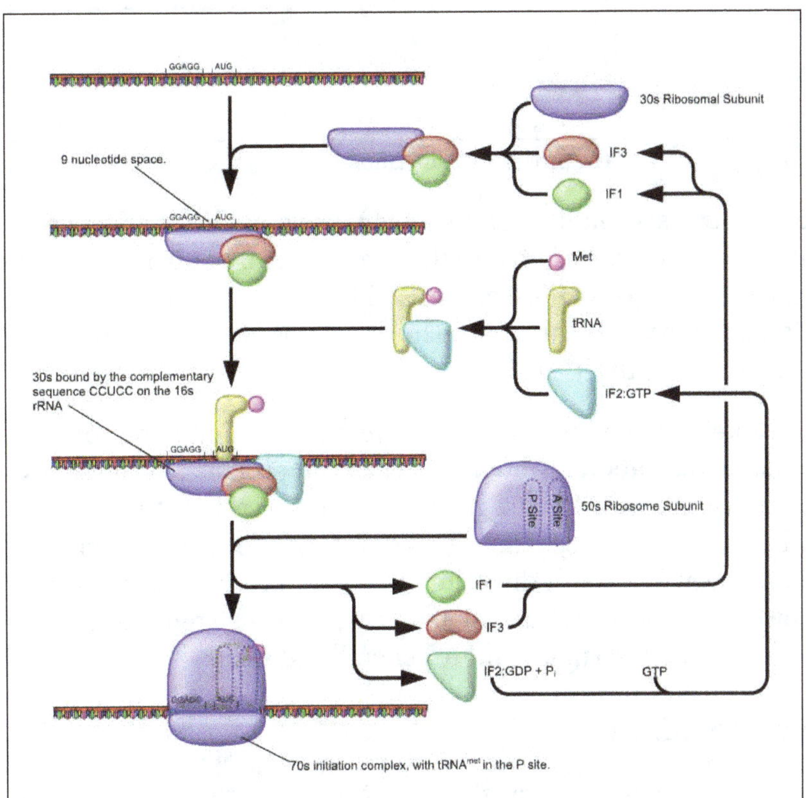

The process of translation in prokaryotes.

Eukaryotes

Pre-initiation Complex Regulation

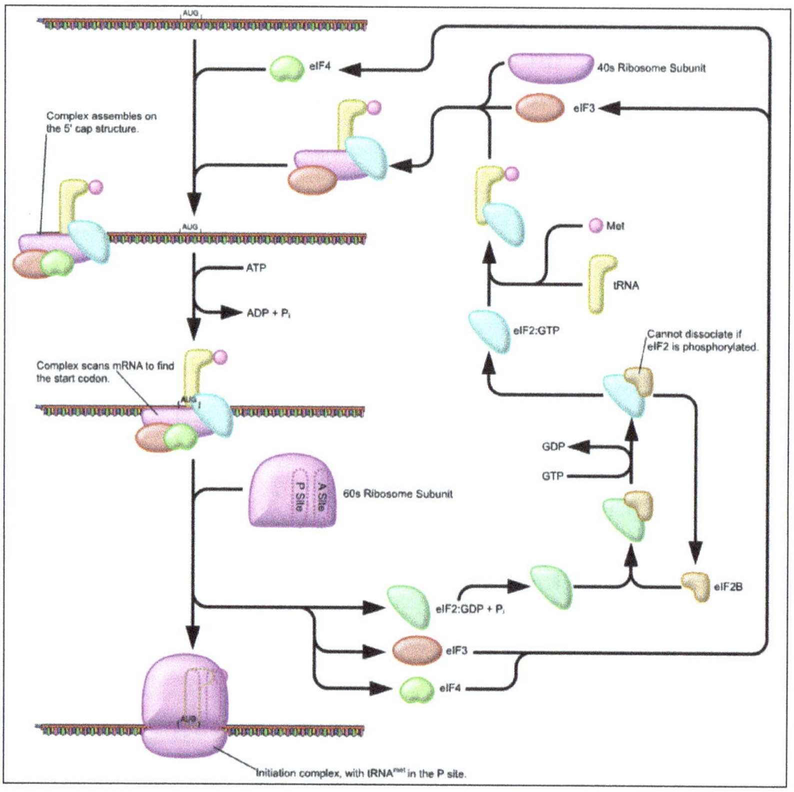

The process of translation in eukaryotes.

The regulation of translation in eukaryotes is more complex than in prokaryotes. Initially, the eIF4F complex is recruited to the 5′ cap, which in turn recruits the ribosomal complex to the 5′ UTR. Both eIF4E and eIF4G bind the 5′ UTR, which limit the rate at which translational initiation can occur. However, this is not the only regulatory step of translation that involves the 5′ UTR.

RNA-binding proteins sometimes serve to prevent the pre-initiation complex from forming. An example is regulation of the *msl2* gene. The protein SXL attaches to an intron segment located within the 5′ UTR segment of the primary transcript, which leads to the inclusion of the intron after processing. This sequence allows the recruitment of proteins that bind simultaneously to both the 5′ and 3′ UTR, not allowing translation proteins to assemble. However, it has also been noted that SXL can also repress translation of RNAs that do not contain a poly(A) tail, or more generally, 3′ UTR.

Closed-loop Regulation

Another important regulator of translation is the interaction between 3′ UTR and the 5′ UTR.

The closed-loop structure inhibits translation. This has been observed in *Xenopus laevis* in which eIF4E bound to the 5′ cap interacts with Maskin bound to CPEB on the 3′ UTR creating translationally inactive transcripts. This translational inhibition is lifted once CPEB is phosphorylated, displacing the Maskin binding site, allowing for the polymerization of the PolyA tail, which can recruit the translational machinery by means of PABP. However, it is important to note that this mechanism has been under great scrutiny.

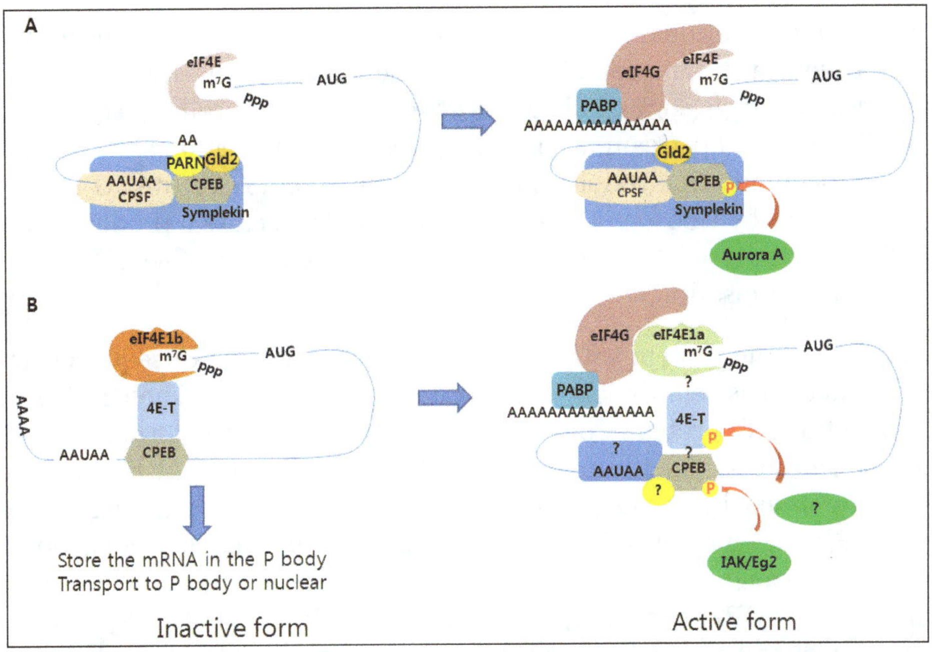

The various forms of mRNA and how each affects translational regulation.

Ferritin Regulation

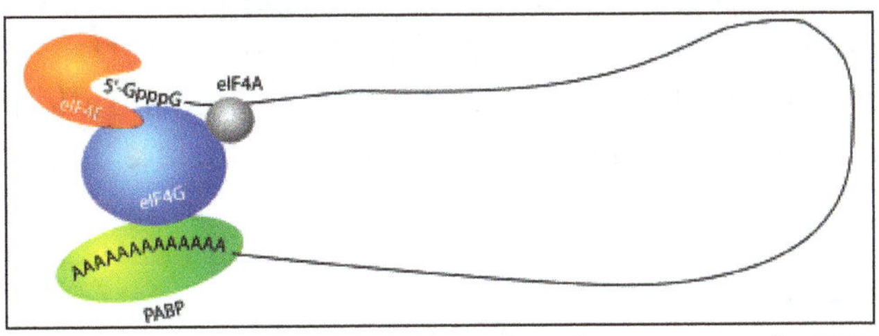

Interactions between proteins bound to the 3′ UTR and 5′ UTR causing a circularization that regulates translation.

Iron levels in cells are maintained by translation regulation of many proteins involved in iron storage and metabolism. The 5′ UTR has the ability to form a hairpin loop secondary structure (known as the IRE, Iron response element) that is recognized by iron-regulatory proteins (IRP1 and IRP2). In low levels of iron, the ORF of the target mRNA is blocked as a result of steric hindrance from the binding of IRP1 and IRP2 to the iron-response element. When iron is high, then the two iron-regulatory proteins do not bind as strongly, and allow proteins to be expressed that have a role in iron concentration control. This function has gained some interest after it was revealed that the translation of amyloid precursor protein may be disrupted due to a single-nucleotide polymorphism to the iron response element found in the 5′ UTR of its mRNA, leading to a spontaneous increased risk of Alzheimer's Disease.

uORFs and Reinitiation

Another form of translational regulation in eukaryotes comes from unique elements on the 5′ UTR called Upstream Open Reading Frames (uORF). These elements are fairly common, occurring in 35-49% of all human genes. A uORF is a coding sequence located in the 5′ UTR located upstream of the coding sequences initiation site. These uORFs contain their own initiation codon, known as an upstream AUG (uAUG). This codon can be scanned for by ribosomes and then translated to create a product, which can regulate the translation of the main protein coding sequence or other uORFs that may exist on the same transcript.

The translation of the protein within the main ORF after a uORF sequence has been translated is known as reinitiation. The process of reinitiation is known to reduce the translation of the ORF protein. Control of protein regulation is determined by the distance between the uORF and the first codon in the main ORF. A uORF has been found to increase reinitiation with the longer distance between its uAUG and the start codon of the main ORF, which indicates that the ribosome needs to reacquire translation factors before it can carry out translation of the main protein. For example, *ATF4* regulation is performed by two uORFs further upstream, named uORF1 and uORF2, which contain three amino acids and fifty-nine amino acids, respectively. The location of uORF2 overlaps with the *ATF4* ORF. During normal conditions, the uORF1 is translated, and then translation of uORF2 occurs only after eIF2-TC has been reacquired. Translation of the uORF2 requires that the ribosomes pass by the *ATF4* ORF, whose start codon is located within uORF2. This leads to its repression. However, during stress conditions, the 40S ribosome will bypass uORF2 because of a decrease in concentration of eIF2-TC, which means the ribosome does not acquire one in time to translate uORF2. Instead *ATF4* is translated.

Other Mechanisms

In addition to reinitiation, uORFs contribute to translation initiation based on:

- The nucleotides of an uORF may code for a codon that leads to a highly structured mRNA, causing the ribosome to stall.
- Cis- and trans- regulation on translation of the main protein coding sequence.
- Interactions with IRES sites.

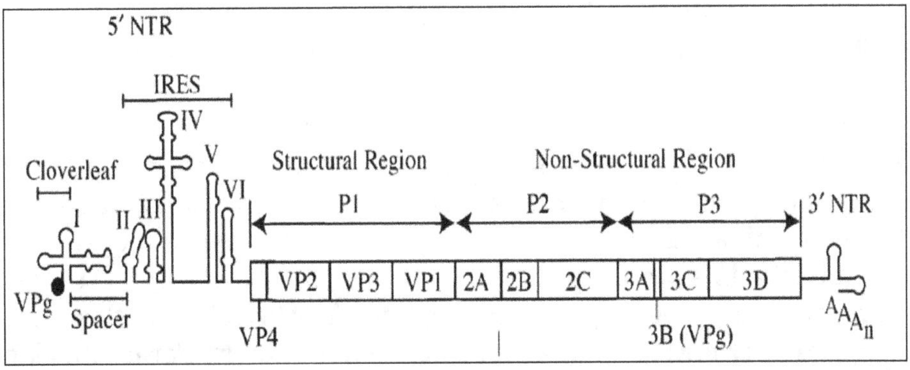

An example IRES in the 5′ UTR of the Poliovirus genome.

Internal Ribosome Entry Sites and Viruses

Viral (as well as some eukaryotic) 5′ UTRs contain internal ribosome entry sites, which is a cap-independent method of translational activation. Instead of building up a complex at the 5′ cap, the IRES allows for direct binding of the ribosomal complexes to the transcript to begin translation. The IRES enables the viral transcript to translate more efficiently due to the lack of needing a preinitation complex, allowing the virus to replicate quickly.

Role in Transcriptional Regulation

msl-2 Transcript

Transcription of the *msl-2* transcript is regulated by multiple binding sites for *Sxl* at the 5′ UTR. In particular, these poly-uracil sites are located close to a small intron that is spliced in males, but kept in females through splicing inhibition. This splicing inhibition is maintained by *Sxl*. When present, *Sxl* will repress the translation of *msl2* by increasing translation of a start codon located in a uORF in the 5′ UTR. Also, *Sxl* outcompetes TIA-1 to a poly(U) region and prevents snRNP (a step in alternative splicing) recruitment to the 5′ splice site.

Three Prime Untranslated Region

In molecular genetics, the three prime untranslated region (3'-UTR) is the section of messenger RNA (mRNA) that immediately follows the translation termination codon. An mRNA molecule is transcribed from the DNA sequence and is later translated into protein. Several regions of the mRNA molecule are not translated into protein including the 5' cap, 5' untranslated region, 3' untranslated region, and the poly(A) tail. The 3'-UTR often contains regulatory regions that post-transcriptionally influence gene expression.

Regulatory regions within the 3'-untranslated region can influence polyadenylation, translation efficiency, localization, and stability of the mRNA. The 3'-UTR contains both binding sites for regulatory proteins as well as microRNAs (miRNAs). By binding to specific sites within the 3'-UTR, miRNAs can decrease gene expression of various mRNAs by either inhibiting translation or directly causing degradation of the transcript. The 3'-UTR also has silencer regions which bind to repressor proteins and will inhibit the expression of the mRNA. Many 3'-UTRs also contain AU-rich elements (AREs). Proteins bind AREs to affect the stability or decay rate of transcripts in a localized manner or affect translation initiation. Furthermore, the 3'-UTR contains the sequence AAUAAA that directs addition of several hundred adenine residues called the poly(A) tail to the end of the mRNA transcript. Poly(A) binding protein (PABP) binds to this tail, contributing to regulation of mRNA translation, stability, and export. For example, poly (A) tail bound PABP interacts with proteins associated with the 5' end of the transcript, causing a circularization of the mRNA that promotes translation. The 3'-UTR can also contain sequences that attract proteins to associate the mRNA with the cytoskeleton, transport it to or from the cell nucleus, or perform other types of localization. In addition to sequences within the 3'-UTR, the physical characteristics of the region, including its length and secondary structure, contribute to translation regulation. These diverse mechanisms of gene regulation ensure that the correct genes are expressed in the correct cells at the appropriate times.

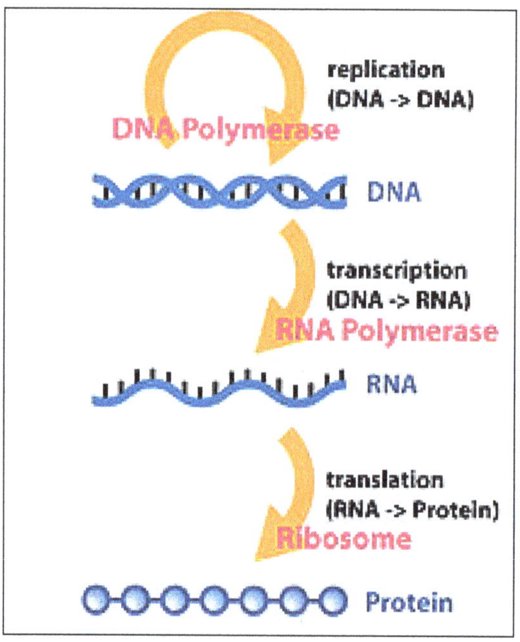

The flow of information within a cell. DNA is first transcribed into RNA, which is subsequently translated into protein.

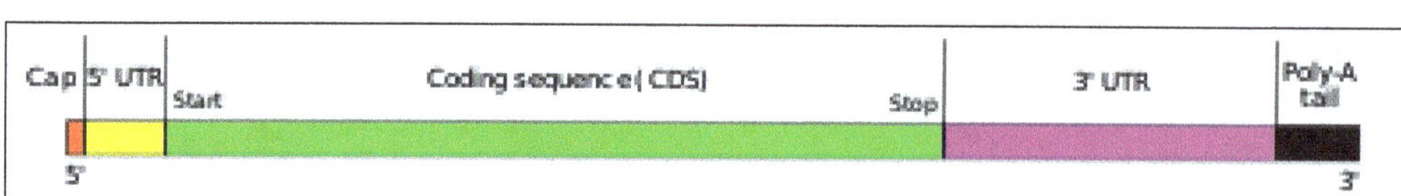

mRNA structure, approximately to scale for a human mRNA, where the median length of 3'UTR is 700 nucleotides.

Physical Characteristics

The 3'-UTR of mRNA has a great variety of regulatory functions that are controlled by the physical characteristics of the region. One such characteristic is the length of the 3'-UTR, which in the mammalian genome has considerable variation. This region of the mRNA transcript can range from 60 nucleotides to about 4000. On average the length for the 3'-UTR in humans is approximately 800 nucleotides, while the average length of 5'-UTRs is only about 200 nucleotides. The length of the 3'-UTR is significant since longer 3'-UTRs are associated with lower levels of gene expression. One possible explanation for this phenomenon is that longer regions have a higher probability of possessing more miRNA binding sites that have the ability to inhibit translation. In addition to length, the nucleotide composition also differs significantly between the 5' and 3'-UTR. The mean G+C percentage of the 5'-UTR in warm-blooded vertebrates is about 60% as compared to only 45% for 3'-UTRs. This is important because an inverse correlation has been observed between the G+C% of 5' and 3'-UTRs and their corresponding lengths. The UTRs that are GC-poor tend to be longer than those located in GC-rich genomic regions.

Sequences within the 3'-UTR also have the ability to degrade or stabilize the mRNA transcript. Modifications that control a transcript's stability allow expression of a gene to be rapidly controlled without altering translation rates. One group of elements in the 3'-UTR that can help destabilize an mRNA transcript are the AU-rich elements (AREs). These elements range in size from 50-150 base pairs and generally contain multiple copies of the pentanucleotide AUUUA. Early studies indicated that AREs can vary in sequence and fall into three main classes that differ in the number and arrangement of motifs. Another set of elements that is present in both the 5' and 3'-UTR are iron response elements (IREs). The IRE is a stem-loop structure within the untranslated regions of mRNAs that encode proteins involved in cellular iron metabolism. The mRNA transcript containing this element is either degraded or stabilized depending upon the binding of specific proteins and the intracellular iron concentrations.

The 3'-UTR also contains sequences that signal additions to be made, either to the transcript itself or to the product of translation. For example, there are two different polyadenylation signals present within the 3'-UTR that signal the addition of the poly(A) tail. These signals initiate the synthesis of the poly(A) tail at a defined length of about 250 base pairs. The primary signal used is the nuclear polyadenylation signal (PAS) with the sequence AAUAAA located toward the end of the 3'-UTR. However, during early development cytoplasmic polyadenylation can occur instead and regulate the translational activation of maternal mRNAs. The element that controls this process is called the CPE which is AU-rich and located in the 3'-UTR as well. The CPE generally has the structure UUUUUUAU and is usually within 100 base pairs of the nuclear PAS. Another specific addition signaled by the 3'-UTR is the incorporation of selenocysteine at UGA codons of mRNAs encoding selenoproteins. Normally the UGA codon encodes for a stop of translation, but in this case a conserved stem-loop structure called the selenocysteine insertion sequence(SECIS) causes for the insertion of selenocysteine instead.

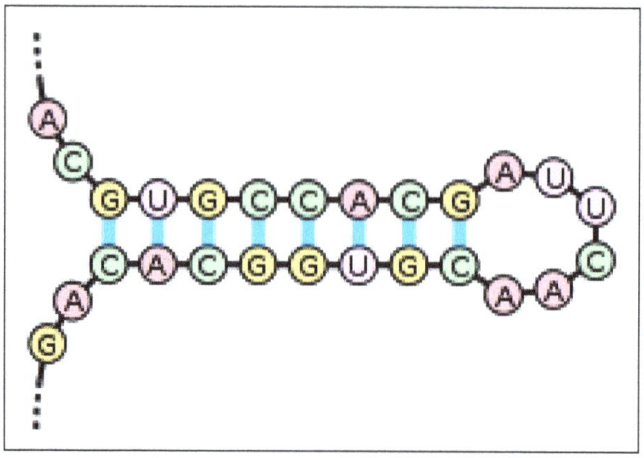

Stem-loop structure of an RNA molecule.

Role in Gene Expression

The 3'-untranslated region plays a crucial role in gene expression by influencing the localization, stability, export, and translation efficiency of an mRNA. It contains various sequences that are involved in gene expression, including microRNA response elements (MREs), AU-rich elements (AREs), and the poly(A) tail. In addition, the structural characteristics of the 3'-UTR as well as its use of alternative polyadenylation play a role in gene expression.

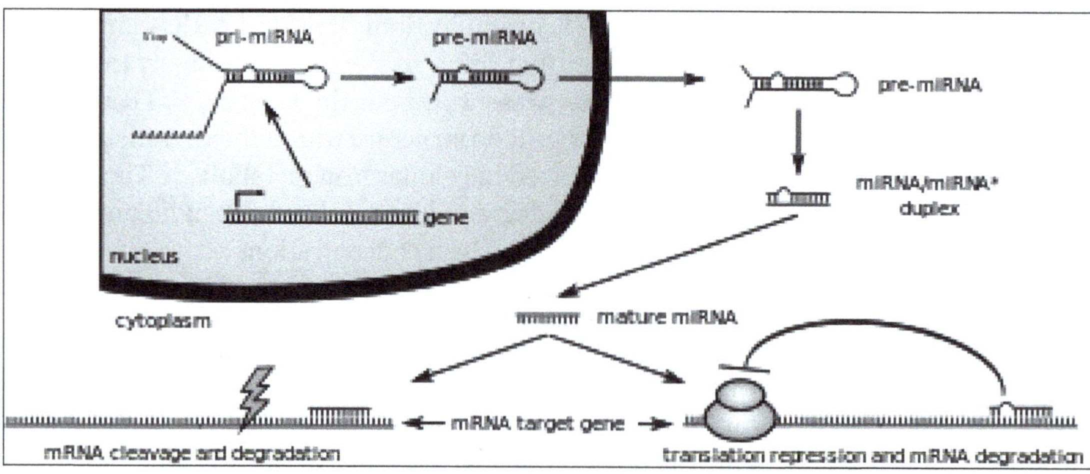

The role of miRNA in gene regulation.

MicroRNA Response Elements

The 3'-UTR often contains microRNA response elements (MREs), which are sequences to which miRNAs bind. miRNAs are short, non-coding RNA molecules capable of binding to mRNA transcripts and regulating their expression. One miRNA mechanism involves partial base pairing of the 5' seed sequence of an miRNA to an MRE within the 3'-UTR of an mRNA; this binding then causes translational repression.

AU-rich Elements

In addition to containing MREs, the 3'-UTR also often contains AU-rich elements (AREs), which are 50 to 150 bp in length and usually include many copies of the sequence AUUUA. ARE binding proteins (ARE-BPs) bind to AU-rich elements in a manner that is dependent upon tissue type, cell type, timing, cellular localization, and environment. In response to different intracellular and extracellular signals, ARE-BPs can promote mRNA decay, affect mRNA stability, or activate translation. This mechanism of gene regulation is involved in cell growth, cellular differentiation, and adaptation to external stimuli. It therefore acts on transcripts encoding cytokines, growth factors, tumor suppressors, proto-oncogenes, cyclins, enzymes, transcription factors, receptors, and membrane proteins.

Poly(A) Tail

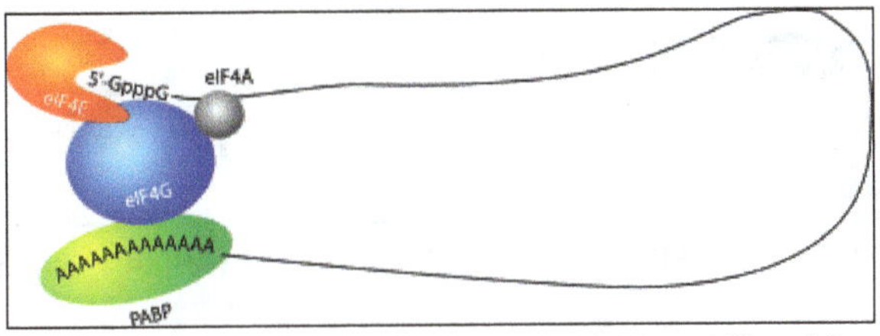

Circularization of the mRNA transcript is mediated by proteins interacting with the 5' cap and poly(A) tail.

The poly(A) tail contains binding sites for poly(A) binding proteins (PABPs). These proteins cooperate with other factors to affect the export, stability, decay, and translation of an mRNA. PABPs bound to the poly(A) tail may also interact with proteins, such as translation initiation factors, that are bound to the 5' cap of the mRNA. This interaction causes circularization of the transcript, which subsequently promotes translation initiation. Furthermore, it allows for efficient translation by causing recycling of ribosomes. While the presence of a poly(A) tail usually aids in triggering translation, the absence or removal of one often leads to exonuclease-mediated degradation of the mRNA. Polyadenylation itself is regulated by sequences within the 3'-UTR of the transcript. These sequences include cytoplasmic polyadenylation elements (CPEs), which are uridine-rich sequences that contribute to both polyadenylation activation and repression. CPE-binding protein (CPEB) binds to CPEs in conjunction with a variety of other proteins in order to elicit different responses.

Structural Characteristics

While the sequence that constitutes the 3'-UTR contributes greatly to gene expression, the structural characteristics of the 3'-UTR also play a large role. In general, longer 3'-UTRs correspond to lower expression rates since they often contain more miRNA and protein binding sites that are involved in inhibiting translation. Human transcripts possess 3'-UTRs that are on average twice as long as other mammalian 3'-UTRs. This trend reflects the high level of complexity involved in human gene regulation. In addition to length, the secondary structure of the 3'-untranslated region also has regulatory functions. Protein factors can either aid or disrupt folding of the region into various secondary structures. The most common structure is a stem-loop, which provides a scaffold for RNA binding proteins and non-coding RNAs that influence expression of the transcript.

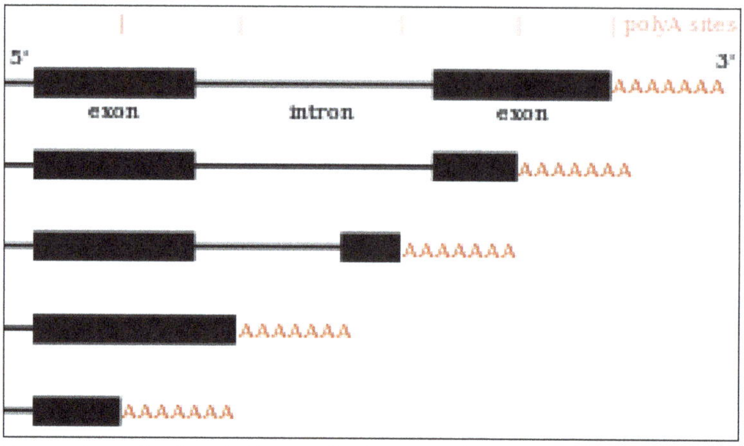

Alternative polyadenylation results in transcripts with different 3'-UTRs.

Alternative Polyadenylation

Another mechanism involving the structure of the 3'-UTR is called alternative polyadenylation (APA), which results in mRNA isoforms that differ only in their 3'-UTRs. This mechanism is especially useful for complex organisms as it provides a means of expressing the same protein but in varying amounts and locations. It is utilized by about half of human genes. APA can result from the presence of multiple polyadenylation sites or mutually exclusive terminal exons. Since it can affect the presence of protein and miRNA binding sites, APA can cause differential expression of mRNA transcripts by influencing their stability, export to the cytoplasm, and translation efficiency.

Methods of Study

Scientists use a number of methods to study the complex structures and functions of the 3' UTR. Even if a given 3'-UTR in an mRNA is shown to be present in a tissue, the effects of localization, functional half-life, translational efficiency, and trans-acting elements must be determined to understand the 3'-UTR's full functionality. Computational approaches, primarily by sequence analysis, have shown the existence of AREs in approximately 5 to 8% of human 3'-UTRs and the presence of one or more miRNA targets in as many as 60% or more of human 3'-UTRs. Software can rapidly compare millions of sequences at once to find similarities between various 3' UTRs within the genome. Experimental approaches have been used to define sequences that associate with specific RNA-binding proteins; specifically, recent improvements in sequencing and cross-linking techniques have enabled fine mapping of protein binding sites within the transcript. Induced site-specific mutations, for example those that affect the termination codon, polyadenylation signal, or secondary structure of the 3'-UTR, can show how mutated regions can cause translation deregulation and disease. These types of transcript-wide methods should help our understanding of known cis elements and trans-regulatory factors within 3'-UTRs.

Disease

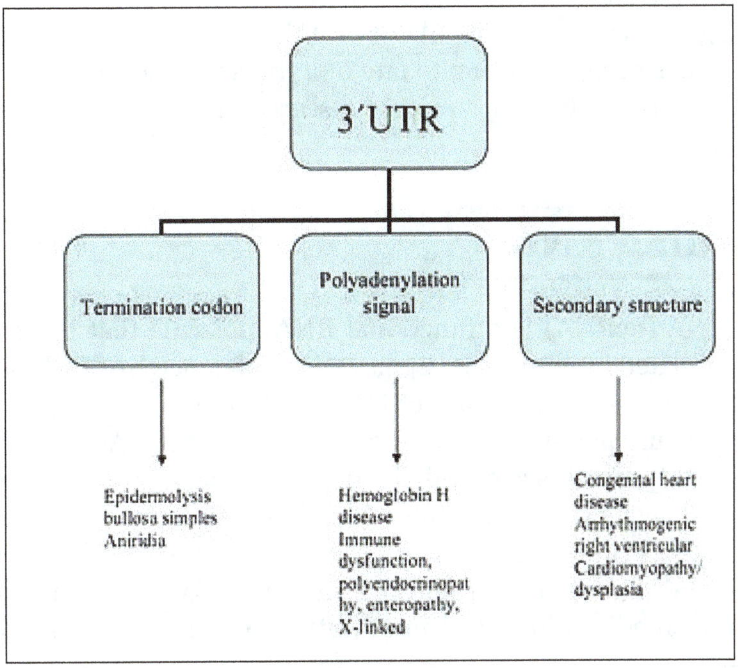

Diseases caused by different mutations within the 3'-UTR.

3'-UTR mutations can be very consequential because one alteration can be responsible for the altered expression of many genes. Transcriptionally, a mutation may affect only the allele and genes that are physically linked. However, since 3'-UTR binding proteins also function in the processing and nuclear export of mRNA, a mutation can also affect other unrelated genes. Dysregulation of ARE-binding proteins (AUBPs) due to mutations in AU-rich regions can lead to diseases including tumorigenesis (cancer), hematopoetic malignancies, leukemogenesis, and developmental delay/autism spectrum disorders. An expanded number of trinucleotide (CTG) repeats in the 3'-UTR of the dystrophia myotonica protein kinase (DMPK) gene causes myotonic dystrophy. Retro-transposal 3-kilobase insertion of tandem repeat sequences within the 3'-UTR of fukutin protein is linked to Fukuyama-type congenital muscular dystrophy. Elements in the 3'-UTR have also been linked to human acute myeloid leukemia, alpha-thalassemia, neuroblastoma, Keratinopathy, Aniridia, IPEX syndrome, and congenital heart defects. The few UTR-mediated diseases identified only hint at the countless links yet to be discovered.

Future Development

Despite our current understanding of 3'-UTRs, they are still relative mysteries. Since mRNAs usually contain several overlapping control elements, it is often difficult to specify the identity and function of each 3'-UTR element, let alone the regulatory factors that may bind at these sites. Additionally, each 3'-UTR contains many alternative AU-rich elements and polyadenylation signals. These cis- and trans-acting elements, along with miRNAs, offer a virtually limitless range of control possibilities within a single mRNA. Future research through the increased use of deep-sequencing based ribosome profiling will reveal more regulatory subtleties as well as new control elements and AUBPs. Furthermore, the ultimate fate of a transcript lies in the signal transduction pathway that it is involved in, so future research in this area appears promising.

Non-coding RNA

A non-coding RNA (ncRNA) is a functional RNA molecule that is transcribed from DNA but not translated into proteins. Epigenetic related ncRNAs include miRNA, siRNA, piRNA and lncRNA. In general, ncRNAs function to regulate gene expression at the transcriptional and post-transcriptional level. Those ncRNAs that appear to be involved in epigenetic processes can be divided into two main groups; the short ncRNAs (<30 nts) and the long ncRNAs (>200 nts). The three major classes of short non-coding RNAs are microRNAs (miRNAs), short interfering RNAs (siRNAs), and piwi-interacting RNAs (piRNAs). Both major groups are shown to play a role in heterochromatin formation, histone modification, DNA methylation targeting, and gene silencing.

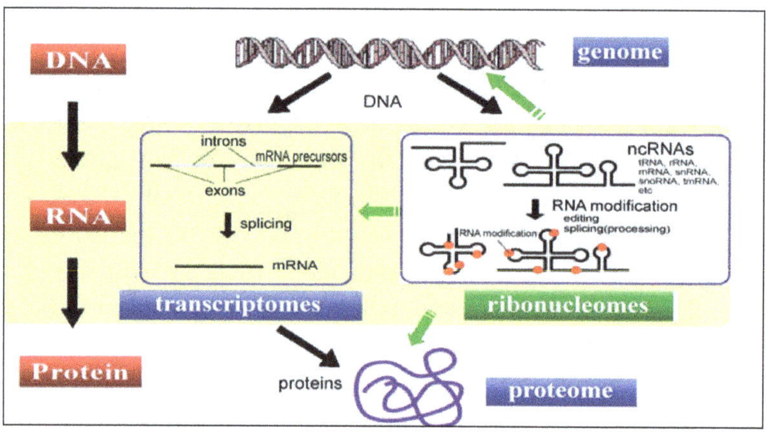

In vivo function of NcRNA in living organisms.

miRNA (microRNA)

Mature microRNAs (miRNAs) are a class of naturally occurring, small non-coding RNA molecules, about 21–25 nucleotides in length. MicroRNAs are partially complementary to one or more messenger RNA (mRNA) molecules, and their main function is to downregulate gene expression in a variety of manners, including translational repression, mRNA cleavage, and deadenylation. They were first described in 1993 by Lee and colleagues, and the term microRNA was coined in 2001. Thousands of miRNAs have since been identified in various organisms through random cloning and sequencing or computational prediction. The miRBase, hosted by the Sanger Institute provides miRNA nomenclature, sequence data, annotation and target prediction information.

miRNA Biogenesis

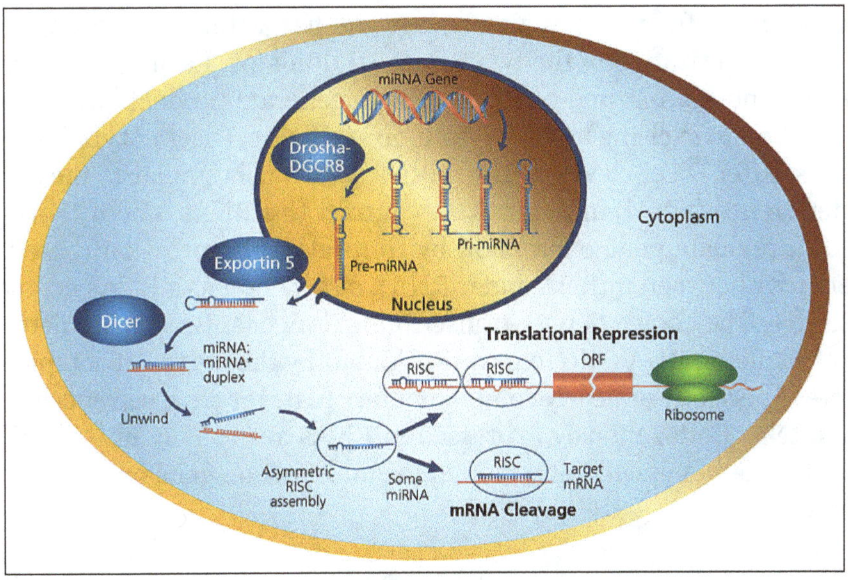

miRNA Pathway.

The genes encoding miRNAs are much longer than the processed mature miRNA molecule. Many miRNAs are known to reside in introns of their pre-mRNA host genes and share their regulatory elements, primary transcript, and have a similar expression profile. For the remainder of miRNA genes that are transcribed from their own promoters, few primary transcripts have been fully identified. MicroRNAs are transcribed by RNA polymerase II as large RNA precursors called pri-miRNAs and comprise of a 5' cap and poly-A tail. The pri-miRNAs are processed in the nucleus by the microprocessor complex, consisting of the RNase III enzyme Drosha4, and the double-stranded-RNA-binding protein, Pasha/DGCR8. The resulting pre-miRNAs are aproximately 70-nucleotides in length and are folded into imperfect stem-loop structures. The pre-miRNAs are then exported into the cytoplasm by the karyopherin exportin 5 (Exp5) and Ran-GTP complex. Ran (ras-related nuclear protein) is a small GTP binding protein belonging to the RAS superfamily that is essential for the translocation of RNA and proteins through the nuclear pore complex7. The Ran GTPase binds Exp5 and forms a nuclear heterotrimer with pre-miRNAs. Once in the cytoplasm, the pre-miRNAs undergo an additional processing step by the RNAse III enzyme Dicer9 generating the miRNA, a double-stranded

RNA aproximately 22 nucleotides in length. Dicer also initiates the formation of the RNA-induced silencing complex (RISC)rev. in . RISC is responsible for the gene silencing observed due to miRNA expression and RNA interference.

Dicer Function

Dicers are large 200 kDa proteins containing an ATPase/RNA helicase, a DUF283 (Domain of unknown function) domain, a PAZ (Piwi, Argonaut and Zwille) domain which can bind the characteristic two-base 3' overhangs of mi and siRNA, two catalytic RNase III domains (RIIIa and RIIIb), and a C-terminal double-stranded RNA-binding domain (dsRBD). Dicer functions as a monomer and has a single processing center with intramolecular dimerization of the two RNase III domains. Each RNase domain independently cuts one RNA strand of the duplex to generate products with 2-nt 3' overhangs. In addition to excising miRNAs from pre-miRNAs, Dicer enzymes process dsRNA to siRNAs. After Dicer cleavage, the miRNA pathway is similar to the central steps of RNA interference (RNAi) in animals. In contrast to siRNAs, microRNAs can direct RISC to down-regulate gene expression by translational repression (based on lower complementarity between miRNA and mRNA), or function as siRNAs do and mediate mRNA cleavage. The choice of post-transcriptional mechanisms is not determined by the origin of silencing RNA (siRNA or miRNA), but by the degree of complementarity. If the miRNA is perfectly or nearly complementary to its target, it can specifically cleave the target mRNA. Endogenously expressed miRNAs are usually imperfectly complementary to their target gene(s) and modulates the effect on gene expression via translational repression.

RISC Complex

When Dicer cleaves the pre-miRNA stem-loop, two complementary short RNA molecules are formed, but only one is integrated into the RISC complex. RISC is a ribonucleoprotein complex containing members of the Argonaute (Ago) family of proteins. Argonaute proteins have endonuclease activity directed against mRNA strands that are complementary to their bound miRNA fragment. Argonaut is also partially responsible for selection of the guide strand and destruction of the passenger strand. The incorporated strand is known as the guide strand and is selected by the argonaute protein on the basis of the stability of the 5' end. The remaining strand, known as the passenger strand (*), is degraded as a RISC complex substrate. Selection of the guide strand from the dsRNA appears to be based primarily on the stability of the termini of the two ends of the dsRNA. The strand with lower stability base pairing of the 2–4 nt at the 5' end of the duplex preferentially associates with RISC and thus becomes the active miRNA. After integration into the active RISC complex, miRNAs exert their regulatory effects by binding to imperfect complementary sites within the 3' untranslated regions (UTRs) of their mRNA targets. The formation of the double-stranded RNA, resulting from the binding of the miRNA, leads to translational repression.

miRNA in Research

Although the first miRNA was identified over ten years ago, it is only recently that researchers have begun to understand the scope and diversity of these regulatory molecules. Growing evidence shows that miRNAs exhibit a variety of crucial regulatory functions related to cell growth, development, and differentiation, and are associated with a wide variety of human diseases. Several miRNAs have been linked to cancer and heart disease. Expression analysis studies reveal perturbed miRNA expression in tumors compared to normal tissues. MicroRNAs are deregulated in breast, lung, and colon cancer, and upregulated in Burkitt's and other human B-cell lymphomas. As a consequence, human miRNAs are likely to be highly useful as biomarkers, especially for future cancer diagnostics, and are rapidly emerging as attractive targets for disease intervention. In addition to their link with cancer, microRNAs play an important role in the control of diverse aspects of cardiac function and dysfunction as well. Including myocyte growth, integrity of the ventricular wall, contractility, gene expression, and maintenance of cardiac rhythm. The misexpression of miRNAs has been shown to be necessary and sufficient for multiple forms of heart disease.

miRNA Research Product Offerings

It is quite clear that miRNA research has great promise and may be the backbone of many of the most cutting-edge medical therapies of tomorrow. To keep up with the growing interest in miRNA, we have made a strong commitment to this rapidly developing area of research and is in the process of developing a comprehensive product portfolio to our customers including miRNA isolation, amplification, profiling, and functional analysis. The mirPremier microRNA Isolation Kit complements the already robust MISSION RNAi product line which includes a broad choice of MISSION siRNA, MISSION miRNA mimics and shRNA products and services such as libraries, mRNA detection reagents, antibodies and AQUA peptides for protein level detection.

Piwi-interacting RNA

Piwi-interacting RNA (piRNA) is the largest class of small non-coding RNA molecules expressed in animal cells. piRNAs form RNA-protein complexes through interactions with piwi-subfamily Argonaute proteins. These piRNA complexes are mostly involved in the epigenetic and post-transcriptional silencing of transposons and other repeat-derived transcripts, but can also be involved in the regulation of other genetic elements in germ line cells. piRNAs are mostly created from loci that function as transposon traps and provide an RNA-mediated adaptive immunity against transposon expansions and invasions. They are distinct from microRNA (miRNA) in size (26–31 nt rather than 21–24 nt), lack of sequence conservation, increased complexity, and independence of Dicer for biogenesis at least in animals. (Plant Dcl2 may play a role in rasi/piRNA biogenesis.)

Double-stranded RNAs silencening repeat elements, then known as repeat associated small interfering RNAs (rasiRNAs), were proposed in Drosophilia in 2001. In 2008, it was still unclear how piRNAs are generated, but potential methods had been suggested, and it was certain their biogenesis pathway is distinct from miRNA and siRNA, while rasiRNAs are a piRNA subspecies.

Characteristics

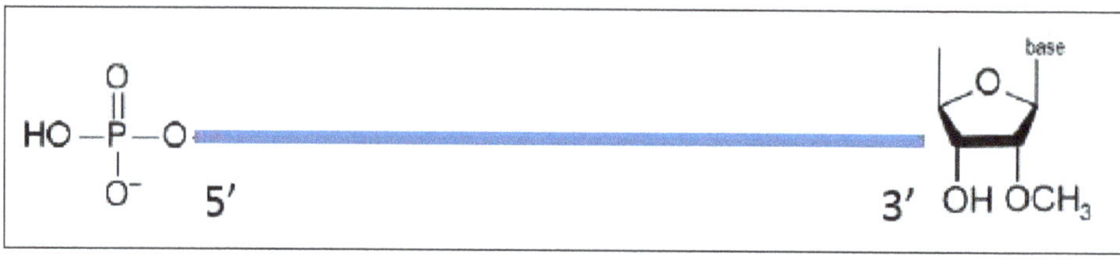

Proposed piRNA structure, with the 3' end 2'-O-methylation.

piRNAs have been identified in both vertebrates and invertebrates, and although biogenesis and modes of action do vary somewhat between species, a number of features are conserved. piRNAs have no clear secondary structure motifs, the length of a piRNA varies between species, from 21 to 31 nucleotides, and the bias for a 5' uridine is common to piRNAs in both vertebrates and invertebrates. piRNAs in *Caenorhabditis elegans* have a 5' monophosphate and a 3' modification that acts to block either the 2' or 3' oxygen, and this has also been confirmed to exist in *Drosophila melanogaster*, zebrafish, mice and rats. This 3' modification is a 2'-O-methylation; the reason for this modification is not clear, but it has been suggested to increase piRNA stability. It is thought that there are many hundreds of thousands of different piRNA species found in mammals. Thus far, over 50,000 unique piRNA sequences have been discovered in mice and more than 13,000 in *D. melanogaster*.

Classification

There are at least three Argonaute subfamilies that have been found in eukaryotes. Unlike the Ago subfamily which is present in animals, plants, and fission yeast, the Piwi subfamily has only been found in animals. RasiRNA has been observed in *Drosophila* and some unicellular eukaryotes but its presence in mammals has not been determined, unlike piRNA which has been observed in many species of invertebrates and vertebrates including mammals; however, since proteins which associate with rasiRNA are found in both vertebrates and invertebrates, it is possible that active rasiRNA exist and have yet to be observed in other animals. RasiRNAs have been observed in *Schizosaccharomyces pombe*, a species of yeast, as well in some plants, neither of which have been observed to contain the Piwi subfamily of Argonaute proteins. It has been observed that both rasiRNA and piRNA are maternally linked, but more specifically it is the Piwi protein subfamily that is maternally linked and therefore leads to the observation that rasiRNA and piRNA are maternally linked.

Biogenesis

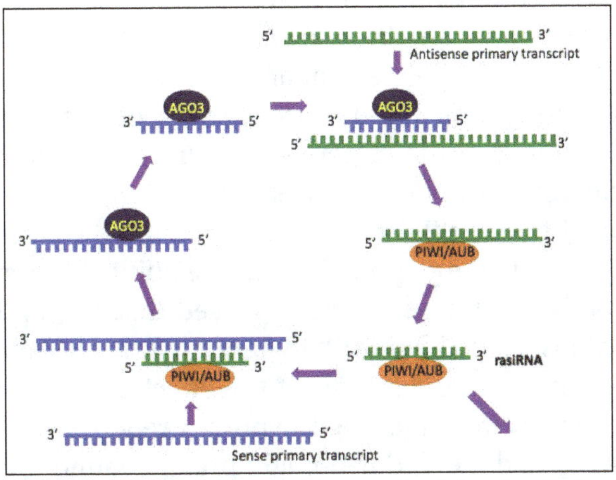

The ping-pong mechanism for the biogenesis of the 5' end of rasiRNA.

The biogenesis of piRNAs is not yet fully understood, although possible mechanisms have been proposed. piRNAs show a significant strand bias, that is, they are derived from one strand of DNA only, and this may indicate that they are the product of long single stranded precursor molecules. A primary processing pathway is suggested to be the only pathway used to produce pachytene piRNAs; in this mechanism, piRNA precursors are transcribed resulting in piRNAs with a tendency to target 5' uridines. Also proposed is a 'Ping Pong' mechanism wherein primary piRNAs recognise their complementary targets and cause the recruitment of piwi proteins. This results in the cleavage of the transcript at a point ten nucleotides from the 5' end of the primary piRNA, producing the secondary piRNA. These secondary piRNAs are targeted toward sequences that possess an adenine at the tenth position. Since the piRNA involved in the ping pong cycle directs its attacks on transposon transcripts, the ping pong cycle acts only at the level of transcription. One or both of these mechanisms may be acting in different species; *C. elegans*, for instance, does have piRNAs, but does not appear to use the ping pong mechanism at all.

A significant number of piRNAs identified in zebrafish and *D. melanogaster* contain adenine at their tenth position, and this has been interpreted as possible evidence of a conserved biosynthetic mechanism across species. Ping-pong signatures have been identified in very primitive animals such as sponges and cnidarians, pointing to the existence of the ping-pong cycle already in the early branches of metazoans.

Ping Pong

The piRNA Ping-Pong pathway was first proposed from studies in *Drosophila* where the piRNA associated with the two cytoplasmic Piwi proteins, Aubergine (Aub) and Argonaute-3 (Ago3) exhibited a high frequency of sequence complementarity over exactly 10 nucleotides at their 5' ends. This relationship is known as the "ping-pong signature" and is also observed in associated piRNA from Mili and Miwi2 proteins isolated from mouse testes. The proposed function of Ping-Pong in *Drosophila* or in mouse remains to be understood, but a leading hypothesis is that the interaction between Aub and Ago3 allows for a cyclic refinement of piRNA that are best suited to target active transposon sequences. Aub piRNA are primarily antisense to transposable element transcripts and are believed to be the main factor in targeting deleterious transcripts through complementarity. Conversely, Ago3 piRNA sequences are predominantly of sense orientation to transposable element transcripts and are derived from the product of Aub cleavage of transposon mRNA. As such, Ago3 piRNA lack the ability to target transposable element transcripts directly. Therefore, it was proposed that Ago3 piRNA guide the production of piRNA that are loaded into Aub by targeting newly exported piRNA cluster transcripts. Several lines of evidence support the effect of Ago3 on the production of Aub piRNA, in particular from examining the piRNA repertoire in *Drosophila* ovaries that are mutant for Ago3 and the Tudor-domain protein Kumo/Qin.

The molecular mechanism that underpins Ping-Pong likely involves several piRNA pathway associated factors. *Qin* was reported to coordinate the loading of Ago3 with piRNA, in addition to interacting with both Aub and Ago3. However, the Tudor protein *krimper* (A1ZAC4) was also shown to interact with both Aub and Ago3 through its Tudor domains while also binding itself through its N-terminal Krimper domain. Specifically, Krimper interacts with Ago3 in its piRNA-unloaded state, while its interaction with Aub is dependent on the symmetrical dimethylation of arginine residues in the N-terminal region of Aub. In Silkmoth germ cells, it was proposed that Vasa protein coordinates the Ping-Pong mechanism of Silkmoth Aub (Siwi) and Ago3. It is likely that the mechanism of Ping-Pong is primarily coordinated by Krimper but factors such as Kumo/Qin and Vasa, in addition to other factors have necessary functions in the Ping-Pong mechanism.

piRNA Phasing

The *Drosophila* piRNA pathway can be separated into two branches: the cytoplasmic branch consisting of Aub and Ago3 operating the Ping-Pong mechanism, and the nuclear branch, pertaining to the co-transcriptional silencing of genomic loci by Piwi in the nucleus. Through complementary strategies, two studies show that Aub and Ago3 target cleavage triggers the 'phased' loading of piRNA into Piwi. Phasing begins with the targeting and cleavage of a complementary target by either Aub or Ago3 associated with a 'responder' piRNA. Once cleaved, the targeted transcript is then processed further by a mechanism believed to require the mitochondrial-associated endonuclease, Zucchini, which leads to the loading of Piwi protein with sequential fragments of the targeted transcript. In this way, the Aub or Ago3 'responder' piRNA sequence cleaves a complementary target that is then sliced at periodic intervals of approximately 27 nucleotides that are sequentially loaded into Piwi protein. Once loaded with piRNA, Piwi then enters the germ cell nucleus to co-transcriptionally silence nascent transcripts with complementarity to its piRNA guide. It is currently unknown whether phasing occurs in other organisms.

Function

The wide variation in piRNA sequences and piwi function over species contributes to the difficulty in establishing the functionality of piRNAs. However, like other small RNAs, piRNAs are thought to be involved in gene silencing, specifically the silencing of transposons. The majority of piRNAs are antisense to transposon sequences, suggesting that transposons are the piRNA target. In mammals it appears that the activity of piRNAs in transposon silencing is most important during the development of the embryo, and in both *C. elegans* and humans, piRNAs are necessary for spermatogenesis.

RNA Silencing

piRNA has a role in RNA silencing via the formation of an RNA-induced silencing complex (RISC). piRNAs interact with piwi proteins that are part of a family of proteins called the Argonautes. These are active in the testes of mammals and are required for germ-cell and stem-cell development in invertebrates. Three piwi subfamily proteins – MIWI, MIWI2 and MILI – have been found to be essential for spermatogenesis in mice. piRNAs direct the piwi proteins to their transposon targets. A decrease or absence of PIWI gene expression is correlated with an increased expression of transposons. Transposons have a high potential to cause deleterious effect on their host, and, in fact, mutations in piRNA pathways are found to reduce fertility in *D. melanogaster*. Further, it is thought that piRNA and endogenous small interfering RNA (endo-siRNA) may have comparable and even redundant functionality in transposon control in mammalian oocytes.

piRNAs appear to affect particular methyltransferases that perform the methylations which are required to recognise and silence transposons, but this relationship is not well understood.

Epigenetic Effects

piRNAs can be transmitted maternally, and based on research in *D. melanogaster*, piRNAs may be involved in maternally derived epigenetic effects. The activity of specific piRNAs in the epigenetic process also requires interactions between piwi proteins and HP1a, as well as other factors.

Accessory Proteins of the piRNA Pathway

Genetic screens examining fertility defects identified a number of proteins that are not Piwi-clade Argonautes, yet produce the same sterility phenotypes as Piwi mutants.

Drosophila Tudor Domain Proteins

Many factors required for the piRNA pathway in *Drosophila* contain Tudor domains that are known to bind symmetrically dimethylated arginine residues (sDMA) present in methylation motifs of Piwi proteins. Piwi proteins are symmetrically dimethylated by the PRMT5 methylosome complex, consisting of Valois (MEP50) and Capsulèen (dart5; PRMT5).

- Tudor (Tud).
- Qin/Kumo.
- Spindle-E (SpnE).
- Krimper.
- Tejas (Tej).
- Vreteno (Vret).
- Papi.
- Yb (*fs(1)Yb*).
- Brother of Yb (BoYB).
- Sister of Yb (SoYB).

Non-tudor Drosophila piRNA Pathway Proteins

- Vasa.
- Maelstrom (Mael).

Drosophila Nuclear piRNA Pathway Proteins

- Rhino (HP1D).
- Deadlock.
- Cutoff.
- SetDB1 (Eggless).
- SuVar3-9.

Investigation

Major advances in the piRNA field have been achieved thanks to the use of next-generation sequencing techniques, such as Solexa, 454, and Illumina platform sequencing.

These techniques allow analysis of highly complex and heterogeneous RNA populations like piRNAs. Due to their small size, expression and amplification of small RNAs can be challenging, so specialised PCR-based methods have been developed in response to this difficulty. However, research also revealed that a number of annotated piRNAs may be false positives, for instance, a majority of piRNAs that were expressed in somatic non-gonadal tissues were considered to derive from non-coding RNA fragments.

RNA Interference

RNA interference (RNAi) is a biological process in which RNA molecules inhibit gene expression or translation, by neutralizing targeted mRNA molecules. Historically, RNAi was known by other names, including co-suppression, post-transcriptional gene silencing (PTGS), and quelling. The detailed study of each of these seemingly different processes elucidated that the identity of these phenomena were all actually RNAi. Andrew Fire and Craig C. Mello shared the 2006 Nobel Prize in Physiology or Medicine for their work on RNA interference in the nematode worm Caenorhabditis elegans, which they published in 1998. Since the discovery of RNAi and its regulatory potentials, it has become evident that RNAi has immense potential in suppression of desired genes. RNAi is now known as precise, efficient, stable and better than antisense technology for gene suppression. However, antisense RNA produced intracellularly by an expression vector may be developed and find utility as novel therapeutic agents.

Two types of small ribonucleic acid (RNA) molecules – microRNA (miRNA) and small interfering RNA (siRNA) – are central to RNA interference. RNAs are the direct products of genes, and these small RNAs can direct enzyme complexes to degrade messenger RNA (mRNA) molecules and thus decrease their activity by preventing translation, via post-transcriptional gene silencing. Moreover, transcription can be inhibited via the pre-transcriptional silencing mechanism of RNA interference, through which an enzyme complex catalyzes DNA methylation at genomic positions complementary to complexed siRNA or miRNA. RNA interference has an important role in defending cells against parasitic nucleotide sequences – viruses and transposons. It also influences development.

The RNAi pathway is found in many eukaryotes, including animals, and is initiated by the enzyme Dicer, which cleaves long double-stranded RNA (dsRNA) molecules into short double-stranded fragments of ~21 nucleotide siRNAs. Each siRNA is unwound into two single-stranded RNAs (ssRNAs), the passenger strand and the guide strand. The passenger strand is degraded and the guide strand is incorporated into the RNA-induced silencing complex (RISC). The most well-studied outcome is post-transcriptional gene silencing, which occurs when the guide strand pairs with a complementary sequence in a messenger RNA molecule and induces cleavage by Argonaute 2 (Ago2), the catalytic component of the RISC. In some organisms, this process spreads systemically, despite the initially limited molar concentrations of siRNA.

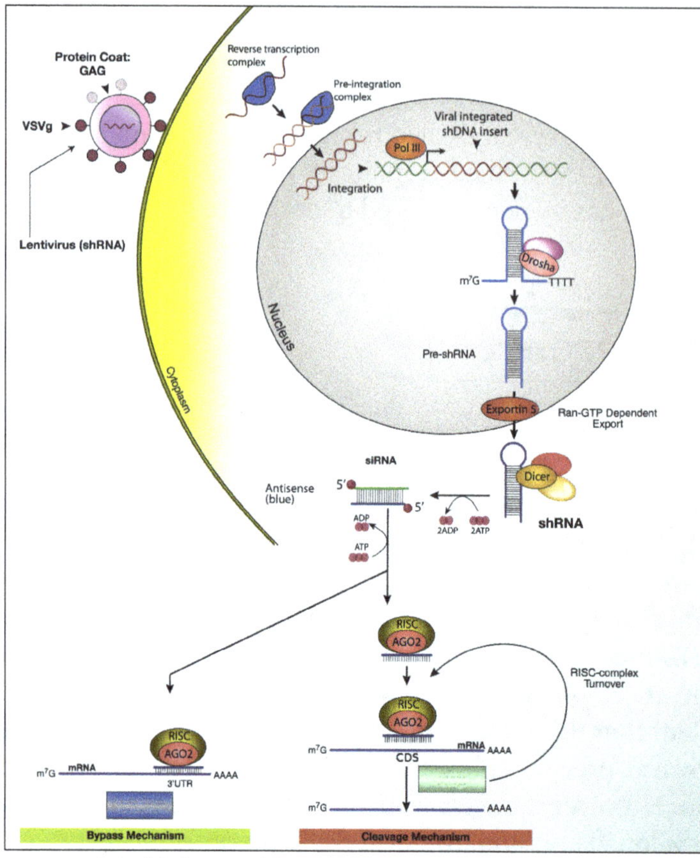

Lentiviral delivery of designed shRNAs and the mechanism of RNA interference in mammalian cells.

RNAi is a valuable research tool, both in cell culture and in living organisms, because synthetic dsRNA introduced into cells can selectively and robustly induce suppression of specific genes of interest. RNAi may be used for large-scale screens that systematically shut down each gene in the cell, which can help to identify the components necessary for a particular cellular process or an event such as cell division. The pathway is also used as a practical tool in biotechnology, medicine and insecticides.

Cellular Mechanism

RNAi is an RNA-dependent gene silencing process that is controlled by the RNA-induced silencing complex (RISC) and is initiated by short double-stranded RNA molecules in a cell's cytoplasm, where they interact with the catalytic RISC component argonaute. When the dsRNA is exogenous (coming from infection by a virus with an RNA genome or laboratory manipulations), the RNA is imported directly into the cytoplasm and cleaved to short fragments by Dicer. The initiating dsRNA can also be endogenous (originating in the cell), as in pre-microRNAs expressed from RNA-coding genes in the genome. The primary transcripts from such genes are first processed to form the characteristic stem-loop structure of pre-miRNA in the nucleus, then exported to the cytoplasm. Thus, the two dsRNA pathways, exogenous and endogenous, converge at the RISC.

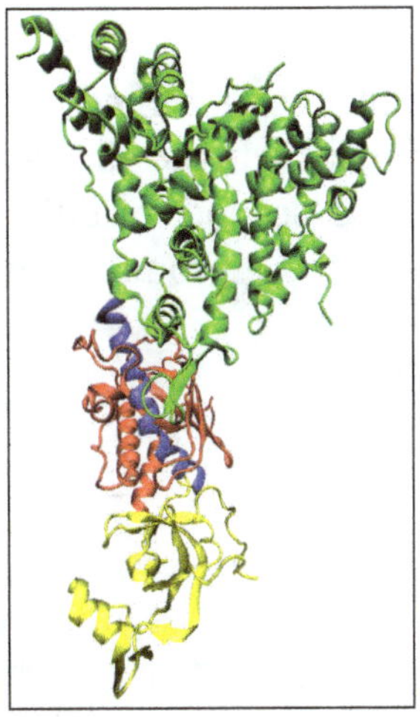

The dicer protein from Giardia intestinalis, which catalyzes the cleavage of dsRNA to siRNAs. The RNase domains are colored green, the PAZ domain yellow, the platform domain red, and the connector helix blue.

Exogenous dsRNA initiates RNAi by activating the ribonuclease protein Dicer, which binds and cleaves double-stranded RNAs (dsRNAs) in plants, or short hairpin RNAs (shRNAs) in humans, to produce double-stranded fragments of 20–25 base pairs with a 2-nucleotide overhang at the 3' end. Bioinformatics studies on the genomes of multiple organisms suggest this length maximizes target-gene specificity and minimizes non-specific effects. These short double-stranded fragments are called small interfering RNAs (siRNAs). These siRNAs are then separated into single strands and integrated into an active RISC, by RISC-Loading Complex (RLC). RLC includes Dicer-2 and R2D2, and is crucial to unite Ago2 and RISC. TATA-binding protein-associated factor 11 (TAF11) assembles the RLC by facilitating Dcr-2-R2D2 tetramerization, which increases the binding affinity to siRNA by 10-fold. Association with TAF11 would convert the R2-D2-Initiator (RDI) complex into the RLC. R2D2 carries tandem double-stranded RNA-binding domains to recognize the thermodynamically stable terminus of siRNA duplexes, whereas Dicer-2 the other less stable extremity. Loading is asymmetric: the MID domain of Ago2 recognizes the thermodynamically stable end of the siRNA. Therefore, the "passenger" (sense) strand whose 5′ end is discarded by MID is ejected, while the saved "guide" (antisense) strand cooperates with AGO to form the RISC.

After integration into the RISC, siRNAs base-pair to their target mRNA and cleave it, thereby preventing it from being used as a translation template. Differently from siRNA, a miRNA-loaded RISC complex scans cytoplasmic mRNAs for potential complementarity. Instead of destructive cleavage (by Ago2), miRNAs rather target the 3′ untranslated region (UTR) regions of mRNAs where they typically bind with imperfect complementarity, thus blocking the access of ribosomes for translation.

Exogenous dsRNA is detected and bound by an effector protein, known as RDE-4 in C. elegans and R2D2 in Drosophila, that stimulates dicer activity. The mechanism producing this length specificity is unknown and this protein only binds long dsRNAs.

In C. elegans this initiation response is amplified through the synthesis of a population of 'secondary' siRNAs during which the dicer-produced initiating or 'primary' siRNAs are used as templates. These 'secondary' siRNAs are structurally distinct from dicer-produced siRNAs and appear to be produced by an RNA-dependent RNA polymerase (RdRP).

MicroRNA

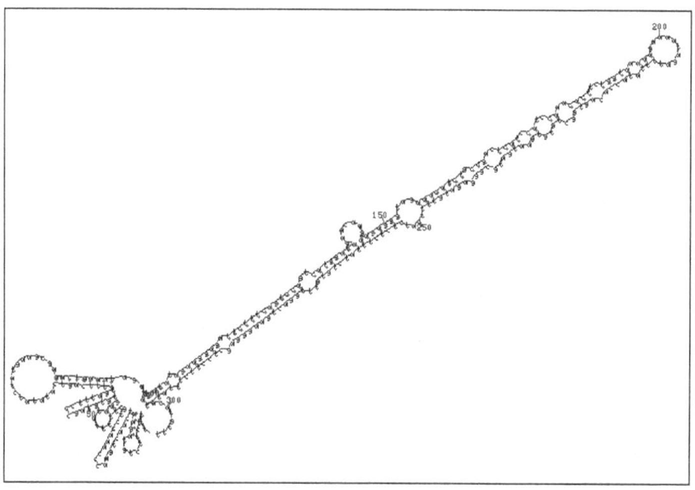

The stem-loop secondary structure of a pre-microRNA from Brassica oleracea.

MicroRNAs (miRNAs) are genomically encoded non-coding RNAs that help regulate gene expression, particularly during development. The phenomenon of RNA interference, broadly defined, includes the endogenously induced gene silencing effects of miRNAs as well as silencing triggered by foreign dsRNA. Mature miRNAs are structurally similar to siRNAs produced from exogenous dsRNA, but before reaching maturity, miRNAs must first undergo extensive post-transcriptional modification. A miRNA is expressed from a much longer RNA-coding gene as a primary transcript known as a pri-miRNA which is processed, in the cell nucleus, to a 70-nucleotide stem-loop structure called a pre-miRNA by the microprocessor complex. This complex consists of an RNase III enzyme called Drosha and a dsRNA-binding protein DGCR8. The dsRNA portion of this pre-miRNA is bound and cleaved by Dicer to produce the mature miRNA molecule that can be integrated into the RISC complex; thus, miRNA and siRNA share the same downstream cellular machinery. First, viral encoded miRNA was described in EBV. Thereafter, an increasing number of microRNAs have been described in viruses. VIRmiRNA is a comprehensive catalogue covering viral microRNA, their targets and anti-viral miRNAs.

siRNAs derived from long dsRNA precursors differ from miRNAs in that miRNAs, especially those in animals, typically have incomplete base pairing to a target and inhibit the translation of many different mRNAs with similar sequences. In contrast, siRNAs typically base-pair perfectly and induce mRNA cleavage only in a single, specific target. In Drosophila and C. elegans, miRNA and siRNA are processed by distinct argonaute proteins and dicer enzymes.

Three Prime Untranslated Regions and MicroRNAs

Three prime untranslated regions (3'UTRs) of messenger RNAs (mRNAs) often contain regulatory sequences that post-transcriptionally cause RNA interference. Such 3'-UTRs often contain both binding sites for microRNAs (miRNAs) as well as for regulatory proteins. By binding to specific sites within the 3'-UTR, miRNAs can decrease gene expression of various mRNAs by either inhibiting translation or directly causing degradation of the transcript. The 3'-UTR also may have silencer regions that bind repressor proteins that inhibit the expression of a mRNA.

The 3'-UTR often contains microRNA response elements (MREs). MREs are sequences to which miRNAs bind. These are prevalent motifs within 3'-UTRs. Among all regulatory motifs within the 3'-UTRs (e.g. including silencer regions), MREs make up about half of the motifs.

As of 2014, the miRBase web site, an archive of miRNA sequences and annotations, listed 28,645 entries in 233 biologic species. Of these, 1,881 miRNAs were in annotated human miRNA loci. miRNAs were predicted to have an average of about four hundred target mRNAs (affecting expression of several hundred genes). Friedman et al. estimate that >45,000 miRNA target sites within human mRNA 3'UTRs are conserved above background levels, and >60% of human protein-coding genes have been under selective pressure to maintain pairing to miRNAs.

Direct experiments show that a single miRNA can reduce the stability of hundreds of unique mRNAs. Other experiments show that a single miRNA may repress the production of hundreds of proteins, but that this repression often is relatively mild (less than 2-fold).

The effects of miRNA dysregulation of gene expression seem to be important in cancer. For instance, in gastrointestinal cancers, nine miRNAs have been identified as epigenetically altered and effective in down regulating DNA repair enzymes.

The effects of miRNA dysregulation of gene expression also seem to be important in neuropsychiatric disorders, such as schizophrenia, bipolar disorder, major depression, Parkinson's disease, Alzheimer's disease and autism spectrum disorders.

RISC Activation and Catalysis

Exogenous dsRNA is detected and bound by an effector protein, known as RDE-4 in

C. elegans and R2D2 in Drosophila, that stimulates dicer activity. This protein only binds long dsRNAs, but the mechanism producing this length specificity is unknown. This RNA-binding protein then facilitates the transfer of cleaved siRNAs to the RISC complex.

In C. elegans this initiation response is amplified through the synthesis of a population of 'secondary' siRNAs during which the dicer-produced initiating or 'primary' siRNAs are used as templates. These 'secondary' siRNAs are structurally distinct from dicer-produced siRNAs and appear to be produced by an RNA-dependent RNA polymerase (RdRP).

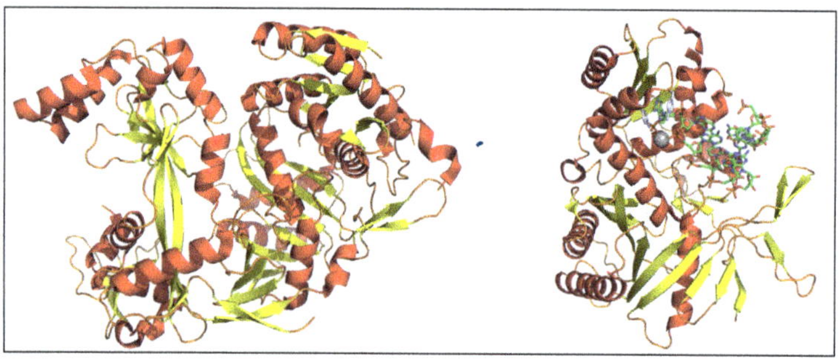

Left: A full-length argonaute protein from the archaea species Pyrococcus furiosus.
Right: The PIWI domain of an argonaute protein in complex with double-stranded RNA.

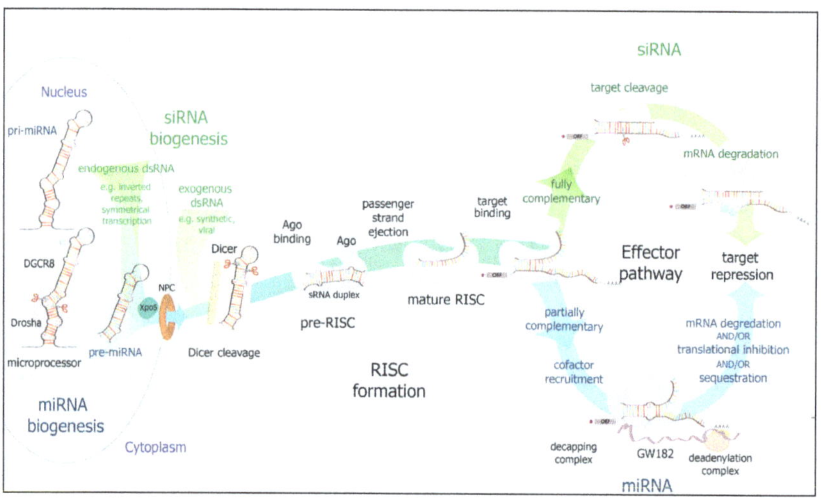

Small RNA Biogenesis: primary miRNAs (pri-miRNAs) are transcribed in the nucleus and fold back onto themselves as hairpins that are then trimmed in the nucleus by a microprocessor complex to form a ~60-70nt hairpin pre-RNA. This pre-miRNA is transported through the nuclear pore complex (NPC) into the cytoplasm, where Dicer further trims it to a ~20nt miRNA duplex (pre-siRNAs also enter the pathway at this step). This duplex is then loaded into Ago to form the "pre-RISC(RNA induced silencing complex)" and the passenger strand is released to form active RISC.

The active components of an RNA-induced silencing complex (RISC) are endonucleases called argonaute proteins, which cleave the target mRNA strand complementary to their bound siRNA. As the fragments produced by dicer are double-stranded, they could each in theory produce a functional siRNA. However, only one of the two strands, which is known as the guide strand, binds the argonaute protein and directs gene silencing. The other anti-guide strand or passenger strand is degraded during RISC activation. Although it was first believed that an ATP-dependent helicase separated these two strands, the process proved to be ATP-independent and performed directly by the protein components of RISC. However, an in vitro kinetic analysis of RNAi in the presence and absence of ATP showed that ATP may be required to unwind and remove the cleaved mRNA strand from the RISC complex after catalysis. The guide strand tends to be the one whose 5' end is less stably paired to its complement, but strand selection is unaffected by the direction in which dicer cleaves the dsRNA before RISC incorporation. Instead, the R2D2 protein may serve as the differentiating factor by binding the more-stable 5' end of the passenger strand.

The structural basis for binding of RNA to the argonaute protein was examined by X-ray crystallography of the binding domain of an RNA-bound argonaute protein. Here, the phosphorylated 5' end of the RNA strand enters a conserved basic surface pocket and makes contacts through a divalent cation (an atom with two positive charges) such as magnesium and by aromatic stacking (a process that allows more than one atom to share an electron by passing it back and forth) between the 5' nucleotide in the siRNA and a conserved tyrosine residue. This site is thought to form a nucleation site for the binding of the siRNA to its mRNA target. Analysis of the inhibitory effect of mismatches in either the 5' or 3' end of the guide strand has demonstrated that the 5' end of the guide strand is likely responsible for matching and binding the target mRNA, while the 3' end is responsible for physically arranging target mRNA into a cleavage-favorable RISC region.

It is not understood how the activated RISC complex locates complementary mRNAs within the cell. Although the cleavage process has been proposed to be linked to translation, translation of the mRNA target is not essential for RNAi-mediated degradation. Indeed, RNAi may be more effective against mRNA targets that are not translated. Argonaute proteins are localized to specific regions in the cytoplasm called P-bodies (also cytoplasmic bodies or GW bodies), which are regions with high rates of mRNA decay; miRNA activity is also clustered in P-bodies. Disruption of P-bodies decreases the efficiency of RNA interference, suggesting that they are a critical site in the RNAi process.

Transcriptional Silencing

Components of the RNAi pathway are used in many eukaryotes in the maintenance of the organization and structure of their genomes. Modification of histones and associated induction of heterochromatin formation serves to downregulate genes pre-transcriptionally; this process is referred to as RNA-induced transcriptional silencing (RITS), and is carried out by a complex of proteins called the RITS complex. In fission yeast this complex contains argonaute, a chromodomain protein Chp1, and a protein called Tas3 of unknown function. As a consequence, the induction and spread of heterochromatic regions requires the argonaute and RdRP proteins. Indeed, deletion of these genes in the fission yeast S. pombe disrupts histone methylation and centromere formation, causing slow or stalled anaphase during cell division. In some cases, similar processes associated with histone modification have been observed to transcriptionally upregulate genes.

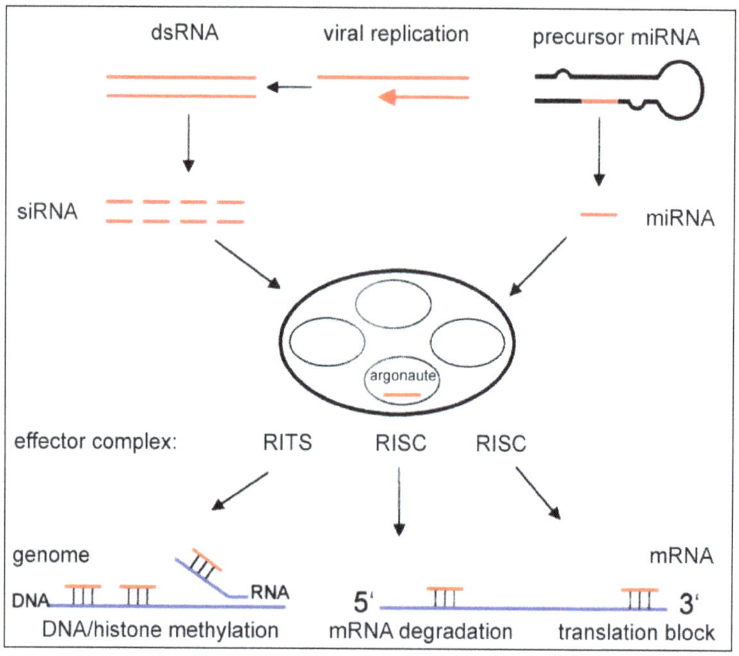

The enzyme dicer trims double stranded RNA, to form small interfering RNA or microRNA. These processed RNAs are incorporated into the RNA-induced silencing complex (RISC), which targets messenger RNA to prevent translation.

The mechanism by which the RITS complex induces heterochromatin formation and organization is not well understood. Most studies have focused on the mating-type region in fission yeast, which may not be representative of activities in other genomic regions/organisms. In maintenance of existing heterochromatin regions, RITS forms a complex with siRNAs complementary to the local genes and stably binds local methylated histones, acting co-transcriptionally to degrade any nascent pre-mRNA transcripts that are initiated by RNA polymerase. The formation of such a heterochromatin region, though not its maintenance, is dicer-dependent, presumably because dicer is required to generate the initial complement of siRNAs that target subsequent transcripts. Heterochromatin maintenance has been suggested to function as a self-reinforcing feedback loop, as new siRNAs are formed from the occasional nascent transcripts by RdRP for incorporation into local RITS complexes. The relevance of observations from fission yeast mating-type regions and centromeres to mammals is not clear, as heterochromatin maintenance in mammalian cells may be independent of the components of the RNAi pathway.

Crosstalk with RNA Editing

The type of RNA editing that is most prevalent in higher eukaryotes converts adenosine nucleotides into inosine in dsRNAs via the enzyme adenosine deaminase (ADAR). It was originally proposed in 2000 that the RNAi and A→I RNA editing pathways might compete for a common dsRNA substrate. Some pre-miRNAs do undergo A→I RNA editing and this mechanism may regulate the processing and expression of mature miRNAs. One mammalian ADAR can sequester siRNAs from RNAi pathway components. Further support for this model comes from studies on ADAR-null C. elegans strains indicating that A→I RNA editing may counteract RNAi silencing of endogenous genes and transgenes.

Variation among Organisms

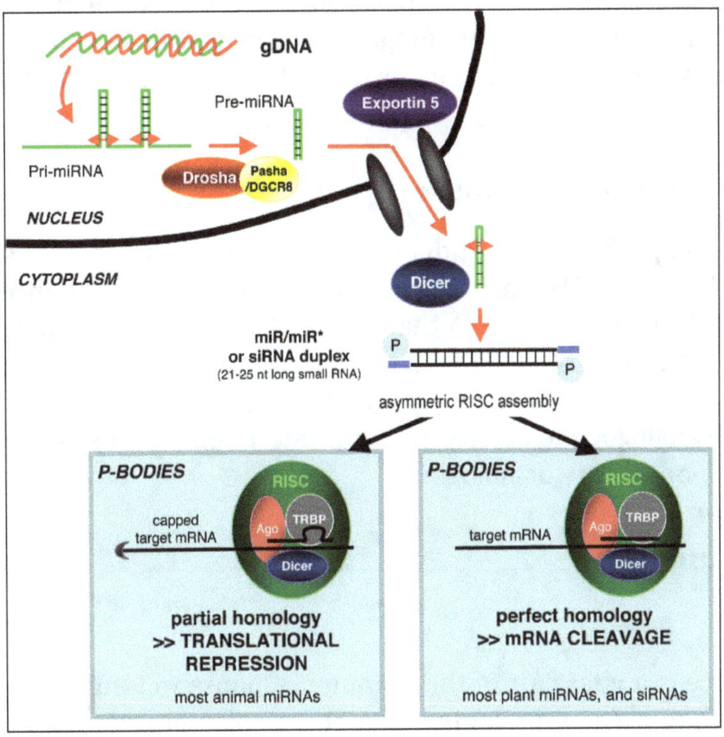

Illustration of the major differences between plant and animal gene silencing. Natively expressed microRNA or exogenous small interfering RNA is processed by dicer and integrated into the RISC complex, which mediates gene silencing.

Organisms vary in their ability to take up foreign dsRNA and use it in the RNAi pathway. The effects of RNA interference can be both systemic and heritable in plants and C. elegans, although not in Drosophila or mammals. In plants, RNAi is thought to propagate by the transfer of siRNAs between cells through plasmodesmata (channels in the cell walls that enable communication and transport). Heritability comes from methylation of promoters targeted by RNAi; the new methylation pattern is copied in each new generation of the cell. A broad general distinction between plants and animals lies in the targeting of endogenously produced miRNAs; in plants, miRNAs are usually perfectly or nearly perfectly complementary to their target genes and induce direct mRNA cleavage by RISC, while animals' miRNAs tend to be more divergent in sequence and induce translational repression. This translational effect may be produced by inhibiting the interactions of translation initiation factors with the messenger RNA's polyadenine tail.

Some eukaryotic protozoa such as Leishmania major and Trypanosoma cruzi lack the RNAi pathway entirely. Most or all of the components are also missing in some fungi, most notably the model organism Saccharomyces cerevisiae. The presence of RNAi in other budding yeast species such as Saccharomyces castellii and Candida albicans, further demonstrates that inducing two RNAi-related proteins from S. castellii facilitates RNAi in S. cerevisiae. That certain ascomycetes and basidiomycetes are missing RNA interference pathways indicates that proteins required for RNA silencing have been lost independently from many fungal lineages, possibly due to the evolution of a novel pathway with similar function, or to the lack of selective advantage in certain niches.

Related Prokaryotic Systems

Gene expression in prokaryotes is influenced by an RNA-based system similar in some respects to RNAi. Here, RNA-encoding genes control mRNA abundance or translation by producing a complementary RNA that anneals to an mRNA. However these regulatory RNAs are not generally considered to be analogous to miRNAs because the dicer enzyme is not involved. It has been suggested that CRISPR interference systems in prokaryotes are analogous to eukaryotic RNA interference systems, although none of the protein components are orthologous.

Biological Functions

Immunity

RNA interference is a vital part of the immune response to viruses and other foreign genetic material, especially in plants where it may also prevent the self-propagation of transposons. Plants such as Arabidopsis thaliana express multiple dicer homologs that are specialized to react differently when the plant is exposed to different viruses. Even before the RNAi pathway was fully understood, it was known that induced gene silencing in plants could spread throughout the plant in a systemic effect and could be transferred from stock to scion plants via grafting. This phenomenon has since been recognized as a feature of the plant adaptive immune system and allows the entire plant to respond to a virus after an initial localized encounter. In response, many plant viruses have evolved elaborate mechanisms to suppress the RNAi response. These include viral proteins that bind short double-stranded RNA fragments with single-stranded overhang ends, such as those produced by dicer. Some plant genomes also express endogenous siRNAs in response to infection by specific types of bacteria. These effects may be part of a generalized response to pathogens that downregulates any metabolic process in the host that aids the infection process.

Although animals generally express fewer variants of the dicer enzyme than plants, RNAi in some animals produces an antiviral response. In both juvenile and adult Drosophila, RNA interference is important in antiviral innate immunity and is active against pathogens such as Drosophila X virus. A similar role in immunity may operate in C. elegans, as argonaute proteins are upregulated in response to viruses and worms that overexpress components of the RNAi pathway are resistant to viral infection.

The role of RNA interference in mammalian innate immunity is poorly understood, and relatively little data is available. However, the existence of viruses that encode genes able to suppress the RNAi response in mammalian cells may be evidence in favour of an RNAi-dependent mammalian immune response, although this hypothesis has been challenged as poorly substantiated. Evidence for the existence of a functional antiviral RNAi pathway in mammalian cells has been presented.

Other functions for RNAi in mammalian viruses also exist, such as miRNAs expressed by the herpes virus that may act as heterochromatin organization triggers to mediate viral latency.

Downregulation of Genes

Endogenously expressed miRNAs, including both intronic and intergenic miRNAs, are most important in translational repression and in the regulation of development, especially on the timing of morphogenesis and the maintenance of undifferentiated or incompletely differentiated cell types such as stem cells. The role of endogenously expressed miRNA in downregulating gene expression was first described in C. elegans in 1993. In plants this function was discovered when the "JAW microRNA" of Arabidopsis was shown to be involved in the regulation of several genes that control plant shape. In plants, the majority of genes regulated by miRNAs are transcription factors; thus miRNA activity is particularly wide-ranging and regulates entire gene networks during development by modulating the expression of key regulatory genes, including transcription factors as well as F-box proteins. In many organisms, including humans, miRNAs are linked to the formation of tumors and dysregulation of the cell cycle. Here, miRNAs can function as both oncogenes and tumor suppressors.

Upregulation of Genes

RNA sequences (siRNA and miRNA) that are complementary to parts of a promoter can increase gene transcription, a phenomenon dubbed RNA activation. Part of the mechanism for how these RNA upregulate genes is known: dicer and argonaute are involved, possibly via histone demethylation. miRNAs have been proposed to upregulate their target genes upon cell cycle arrest, via unknown mechanisms.

RNA Splicing

RNA splicing, in molecular biology, is a form of RNA processing in which a newly made precursor messenger RNA (pre-mRNA) transcript is transformed into a mature messenger RNA (mRNA). During splicing, introns (Non-coding regions) are removed and exons (Coding Regions) are joined together. For nuclear-encoded genes, splicing takes place within the nucleus either during or immediately after transcription. For those eukaryotic genes that contain introns, splicing is usually required in order to create an mRNA molecule that can be translated into protein. For many eukaryotic introns, splicing is carried out in a series of reactions which are catalyzed by the spliceosome, a complex of small nuclear ribonucleo proteins (snRNPs). Self-splicing introns, or ribozymes capable of catalyzing their own excision from their parent RNA molecule, also exist.

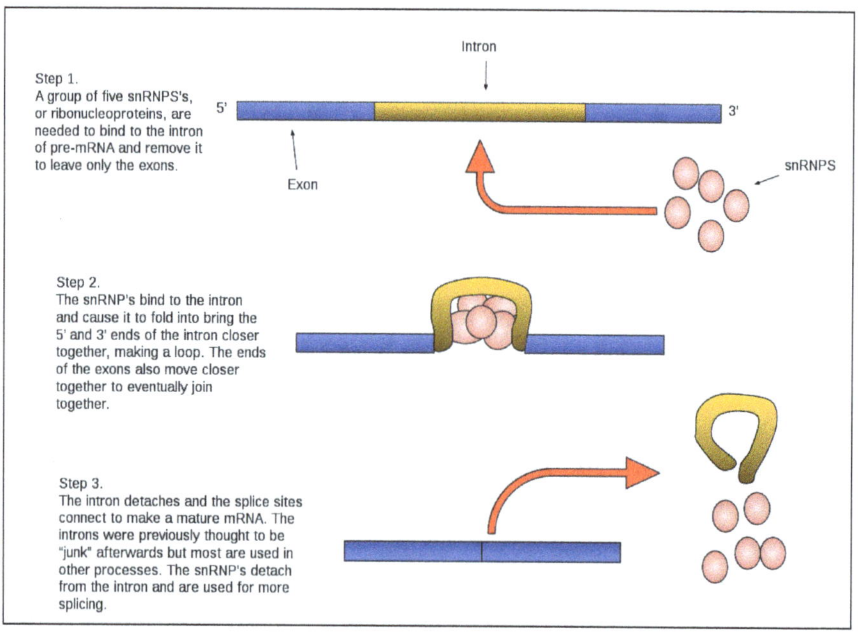

Process of RNA splicing.

Splicing Pathways

Several methods of RNA splicing occur in nature; the type of splicing depends on the structure of the spliced intron and the catalysts required for splicing to occur.

Spliceosomal Complex

Introns

The word *intron* is derived from the terms *intragenic region*, and *intracistron*, that is, a segment of DNA that is located between two exons of a gene. The term intron refers to both the DNA sequence within a gene and the corresponding sequence in the unprocessed RNA transcript. As part of the RNA processing pathway, introns are removed by RNA splicing either shortly after or concurrent with transcription. Introns are found in the genes of most organisms and many viruses. They can be located in a wide range of genes, including those that generate proteins, ribosomal RNA (rRNA), and transfer RNA (tRNA).

Within introns, a donor site (5' end of the intron), a branch site (near the 3' end of the intron) and an acceptor site (3' end of the intron) are required for splicing. The splice donor site includes an almost invariant sequence GU at the 5' end of the intron, within a larger, less highly conserved region. The splice acceptor site at the 3' end of the intron terminates the intron with an almost invariant AG sequence. Upstream (5'-ward) from the AG there is a region high in pyrimidines (C and U), or polypyrimidine tract. Further upstream from the polypyrimidine tract is the branchpoint, which includes an adenine nucleotide involved in lariat formation. The consensus sequence for an intron (in IUPAC nucleic acid notation) is: G-G-[cut]-G-U-R-A-G-U (donor site) intron sequence Y-U-R-A-C (branch sequence 20-50 nucleotides upstream of acceptor site) Y-rich-N-C-A-G-[cut]-G (acceptor site). However, it is noted that the specific sequence of intronic splicing elements and the number of nucleotides between the branchpoint and the nearest 3' acceptor site affect splice site selection. Also, point mutations in the underlying DNA or errors during transcription can activate a *cryptic splice site* in part of the transcript that usually is not spliced. This results in a mature messenger RNA with a missing section of an exon. In this way, a point mutation, which might otherwise affect only a single amino acid, can manifest as a deletion or truncation in the final protein.

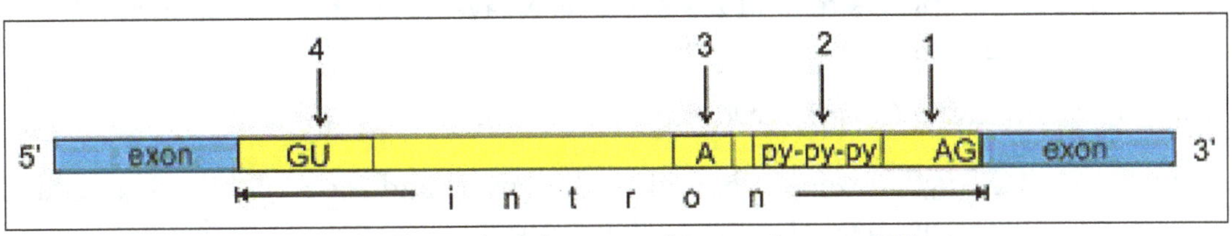

Simple illustration of exons and introns in pre-mRNA.

Formation and Activity

Splicing is catalyzed by the spliceosome, a large RNA-protein complex composed of five small nuclear ribonucleoproteins (snRNPs, pronounced 'snurps'). Assembly and activity of the spliceosome occurs during transcription of the pre-mRNA. The RNA components of snRNPs interact with the intron and are involved in catalysis. Two types of spliceosomes have been identified (major and minor) which contain different snRNPs.

- The major spliceosome splices introns containing GU at the 5' splice site and AG at the 3' splice site. It is composed of the U1, U2, U4, U5, and U6 snRNPs and is active in the nucleus. In addition, a number of proteins including U2 small nuclear RNA auxiliary factor 1 (U2AF35), U2AF2 (U2AF65) and SF1 are required for the assembly of the spliceosome. The spliceosome forms different complexes during the splicing process.

  - 1 binds to the intron branch point sequence.
  - U2AF1 binds at the 3' splice site of the intron.
  - U2AF2 binds to the polypyrimidine tract.

- Complex A (pre-spliceosome).
  - The U2 snRNP displaces SF1 and binds to the branch point sequence and ATP is hydrolyzed.
- Complex B (pre-catalytic spliceosome).
  - The U5/U4/U6 snRNP trimer binds, and the U5 snRNP binds exons at the 5' site, with U6 binding to U2.
- Complex B*.
  - The U1 snRNP is released, U5 shifts from exon to intron, and the U6 binds at the 5' splice site.
- Complex C (catalytic spliceosome).
  - U4 is released, U6/U2 catalyzes transesterification, making the 5'-end of the intron ligate to the A on intron and form a lariat, U5 binds exon at 3' splice site, and the 5' site is cleaved, resulting in the formation of the lariat.
- Complex C* (post-spliceosomal complex).
  - U2/U5/U6 remain bound to the lariat, and the 3' site is cleaved and exons are ligated using ATP hydrolysis. The spliced RNA is released, the lariat is released and degraded, and the snRNPs are recycled.

This type of splicing is termed *canonical splicing* or termed the *lariat pathway*, which accounts for more than 99% of splicing. By contrast, when the intronic flanking sequences do not follow the GU-AG rule, *noncanonical splicing* is said to occur.

- The minor spliceosome is very similar to the major spliceosome, but instead it splices out rare introns with different splice site sequences. While the minor and major spliceosomes contain the same U5 snRNP, the minor spliceosome has different but functionally analogous snRNPs for U1, U2, U4, and U6, which are respectively called U11, U12, U4atac, and U6atac.

Recursive Splicing

In most cases, splicing removes introns as single units from precursor mRNA transcripts. However, in some cases, especially in mRNAs with very long introns, splicing happens in steps, with part of an intron removed and then the remaining intron is spliced out in a following step. This has been found first in the *Ultrabithorax* (*Ubx*) gene of the fruit fly, *Drosophila melanogaster*, and a few other *Drosophila* genes, but cases in human have been reported as well.

Trans-splicing

Trans-splicing is a form of splicing that removes introns or outrons, and joins two exons that are not within the same RNA transcript.

Self-splicing

Self-splicing occurs for rare introns that form a ribozyme, performing the functions of the spliceosome by RNA alone. There are three kinds of self-splicing introns, Group I, Group II and Group III. Group I and II introns perform splicing similar to the spliceosome without requiring any protein. This similarity suggests that Group I and II introns may be evolutionarily related to the spliceosome. Self-splicing may also be very ancient, and may have existed in an RNA world present before protein.

Two transesterifications characterize the mechanism in which group I introns are spliced:

- 3'OH of a free guanine nucleoside (or one located in the intron) or a nucleotide cofactor (GMP, GDP, GTP) attacks phosphate at the 5' splice site.
- 3'OH of the 5' exon becomes a nucleophile and the second transesterification results in the joining of the two exons.

The mechanism in which group II introns are spliced (two transesterification reaction like group I introns) is as follows:

- The 2'OH of a specific adenosine in the intron attacks the 5' splice site, thereby forming the *lariat.*
- The 3'OH of the 5' exon triggers the second transesterification at the 3' splice site, thereby joining the exons together.

tRNA Splicing

tRNA (also tRNA-like) splicing is another rare form of splicing that usually occurs in tRNA. The splicing reaction involves a different biochemistry than the spliceosomal and self-splicing pathways.

In the yeast *Saccharomyces cerevisiae*, a yeast tRNA splicing endonuclease heterotetramer, composed of TSEN54, TSEN2, TSEN34, and TSEN15, cleaves pre-tRNA at two sites in the acceptor loop to form a 5'-half tRNA, terminating at a 2',3'-cyclic phosphodiester group, and a 3'-half tRNA, terminating at a 5'-hydroxyl group, along with a discarded intron. Yeast tRNA kinase then phosphorylates the 5'-hydroxyl group using adenosine triphosphate. Yeast tRNA cyclic phosphodiesterase cleaves the cyclic phosphodiester group to form a 2'-phosphorylated 3' end. Yeast tRNA ligase adds an adenosine monophosphate group to the 5' end of the 3'-half and joins the two halves together. NAD-dependent 2'-phosphotransferase then removes the 2'-phosphate group.

RNA Editing

Occasionally researches encounter a gene with a sequence of nucleotides that does not match exactly that in its RNA product:

- Messenger RNA (mRNA).
- Ribosomal RNA (rRNA).
- Transfer RNA (tRNA).
- MicroRNA (miRNA).

If the product is mRNA, some of the codons in the open reading frame (ORF) of the gene specify different amino acids from those in the protein translated from the mRNA of the gene.

The reason is RNA editing: the alteration of the sequence of nucleotides in the RNA:

- After it has been transcribed from DNA but.
- Before it is translated into protein.

RNA editing occurs by two distinct mechanisms:

- Substitution Editing: Chemical alteration of individual nucleotides (the equivalent of point mutations).
- These alterations are catalyzed by enzymes that recognize a specific target sequence of nucleotides (much like restriction enzymes).
- Cytidine deaminases that convert a C in the RNA to uracil (U).
- Adenosine deaminases that convert an A to inosine (I), which the ribosome translates as a G. Thus a CAG codon (for Gln) can be converted to a CGG codon (for Arg).

- Insertion/Deletion Editing: insertion or deletion of nucleotides in the RNA.
- These alterations are mediated by guide RNA molecules that.
- Base-pair as best they can with the RNA to be edited.
- Serve as a template for the addition (or removal) of nucleotides in the target.

Substitution Editing

The Human APOB Gene

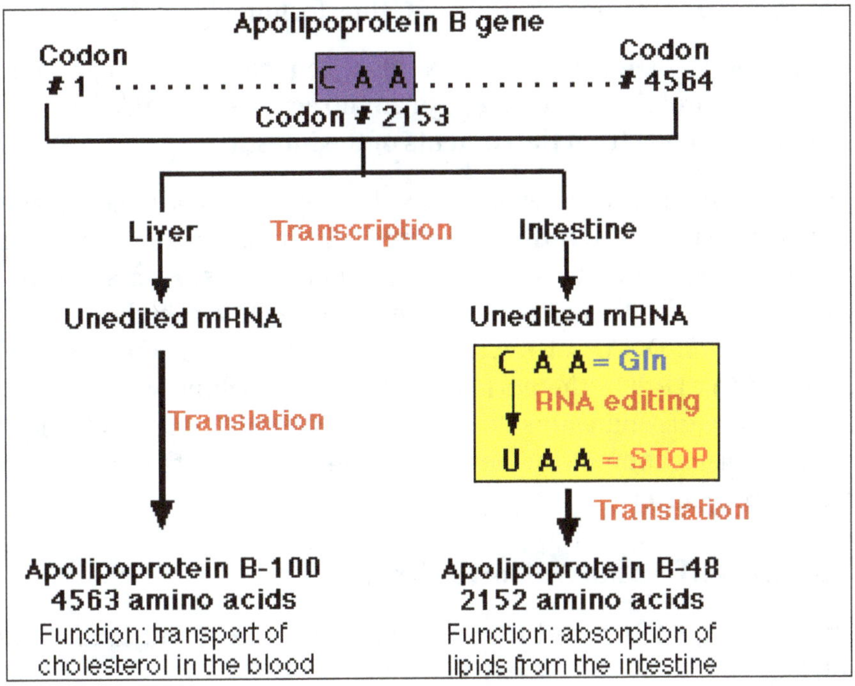

The Human APOB Gene.

Humans have a single locus encoding the APOB gene:

- It contains 29 exons (separated by 28 introns).
- The exons contain a total of 4564 codons.
- Codon 2153 is CAA, which is a codon for the amino acid glutamine (Gln).
- The gene is expressed in cells of both the liver and the intestine.
- In both locations, transcription produces a pre-messenger RNA that must be spliced to produce the mRNA to be translated into protein.
- In the Liver. Here the process occurs normally producing apolipoprotein B-100 — a protein containing 4,563 amino acids — that is essential for the transport of cholesterol and other lipids in the blood.

- In the Intestine.
    - In the cells of the intestine, an additional step of pre-mRNA processing occurs: The chemical modification of the C nucleotide in Codon 2153 (CAA) into a U.
    - This RNA editing changes the codon from one encoding the amino acid glutamine (Gln) to a STOP codon (UAA).
    - The modification is catalyzed by the enzyme cytidine deaminase that:
        - Recognizes the sequence of the RNA at that one place in the molecule and.
        - Catalyzes the deamination of C thus forming U.
    - Translation of the mRNA stops at codon #2153 forming apolipoprotein B-48 — a protein containing 2152 amino acids — that aids in the absorption of dietary lipids from the contents of the intestine.

DNA can also be edited. B cells express another cytidine deaminase (called activation-induced deaminase or AID) that is essential for both class switch recombination (CSR) and somatic hypermutation (SHM) of antibody genes. Humans with disabling mutations in the gene for this enzyme produce only IgM antibodies. However, here the enzyme is acting on DNA, not RNA. In attempting to repair the mismatch formed (dC•dG converted to dU•dG), the normal DNA repair machinery of the cell produces CSR or SHM as the situation warrants. (This process is also responsible for the occasional aberrant translocation of the heavy-chain gene segments to a proto-oncogene. The result is a B-cell cancer — a lymphoma or leukemia.)

Other Examples of Substitution Editing

- Some mRNAs, tRNAs, and rRNAs in both the mitochondria and chloroplasts of plants.
- mRNAs encoding subunits of some receptors of neurotransmitters in the mammalian brain, e.g.
    - The AMPA receptor for Glu.
    - A serotonin receptor.
- A tRNA in the mitochondria of the duckbill platypus.

Insertion/Deletion Editing

The Gene in Mitochondria of Trypanosoma Brucei

Several genes encoded in the mitochondrial DNA of this species (the cause of sleeping sickness in humans) encode transcripts that must be edited to make the mRNA molecules that will be translated into protein.

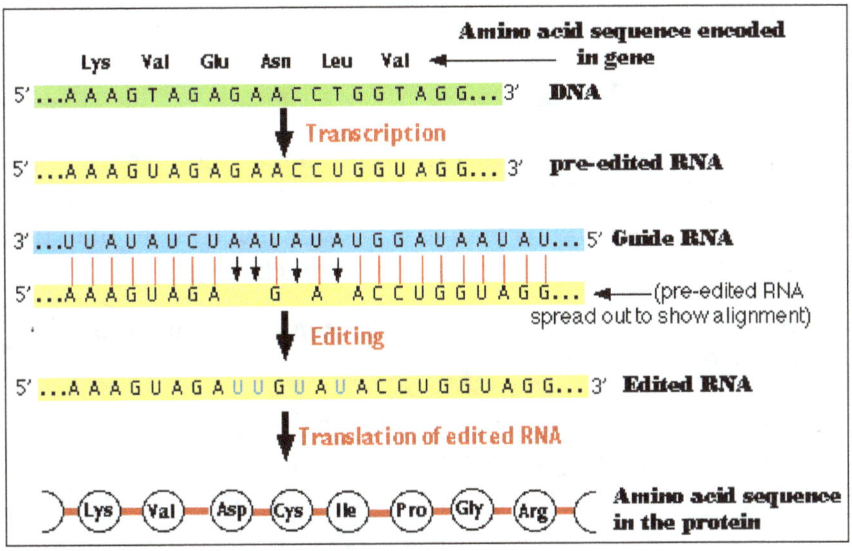

Insertion/Deleting Editing.

Editing requires a special class of RNA molecules called guide RNA (gRNA).

These small molecules have sequences that are complementary to the region around the site to be edited. The guide RNA base-pairs — as best it can — with this region. Note that in addition to the usual purine-pyrimidine pairing of C-G and A-U, G-U base-pairing can also occur.

Because of the lack of precise sequence complementarity, bulges occur either:

- In the guide RNA where, usually, there are As not found in the transcript to be edited.
- In the transcript to be edited.

The bulges are eliminated by cutting the backbone of the shorter molecule and inserting complementary bases.

- In the first case this produces insertions.
- In the second case this produces deletions.

Note that in the example shown here, the insertion of 4 nucleotides has created a frameshift so that the amino acids encoded downstream (after Val) in the edited RNA are entirely different from those specified by the gene itself.

Other Examples of Insertion/Deletion Editing

Insertion/deletion editing has also been found occur with:

- mRNA, rRNA, and tRNA transcripts in the mitochondria of the slime mold Physarum polycephalum.

- In measles virus transcripts.

Importance of RNA Editing

- Perhaps, like alternative splicing, it is a mechanism to increase the number of different proteins available without the need to increase the number of genes in the genome. (The human genome is not that much larger than those of Drosophila and C. elegans, but our proteome is much larger).
- So it can create proteins with slightly different functions to use in specialized circumstances.
    - The ability to synthesize two versions (with different functions) of apolipoprotein B from a single gene is an example. Note that in this case, RNA editing has accomplished the same result as alternative splicing.
    - There is evidence that Drosophila (and humans) use editing to create subtle differences in the properties of.
        - Some voltage-gated ion channels and
        - Some receptors of neurotransmitters in different regions of the brain.
    - In two species of octopus — one tropical and one in the Antarctic — their gene encoding a voltage-gated potassium (K+) channel differs at only four nucleotides, and these have no effect on the electrical properties of the channels. However, the messenger RNAs for the channel are extensively edited in each species to produce channel proteins with different electrical properties. In each case, the channel protein enables rapid firing of action potentials at the temperature of their environment (−1.8°C in the Antarctic and ~30°C in the waters off Puerto Rico).
- RNA (as well as DNA) editing may help protect the genome against.
- Some viruses (and against which they often evolve counter-measures).
- Damage by retrotransposons.
- Other possibilities: The untranslated regions at either end of many messenger RNA (mRNA) molecules — the 5'-UTRs and 3'-UTRs — often contain sequences of nucleotides that permit intramolecular base-pairing resulting in stretches of double-stranded RNA (dsRNA). But dsRNA in the cell runs the risk of being destroyed by RNA interference (RNAi). Perhaps RNA editing in these areas protects against that risk. There is also evidence that RNA editing (converting As to Is in the 3'-UTR) of precursor mRNAs is a signal to retain them within the nucleus ready to be quickly exported if needed by the cell.

- Or perhaps it is simply the legacy of a system that was first used to correct RNA transcripts for harmful mutations in the DNA encoding them. Now with no strong evolutionary pressure to correct the DNA, RNA editing persists.
- So RNA editing appears to be here to stay. In fact, defects in RNA editing are associated with some human cancers as well as with amyotrophic lateral sclerosis (ALS — "Lou Gehrig's disease").

Types of RNA Editing

RNA (ribonucleic acid) editing is a post-transcriptional alteration of RNA sequences and structures via modification, deletion or insertion of nucleotides. This process has been detected in eukaryotes ranging from single-celled protozoa to plants and mammals; as a result, functionally distinctive proteins can be processed from a single gene.

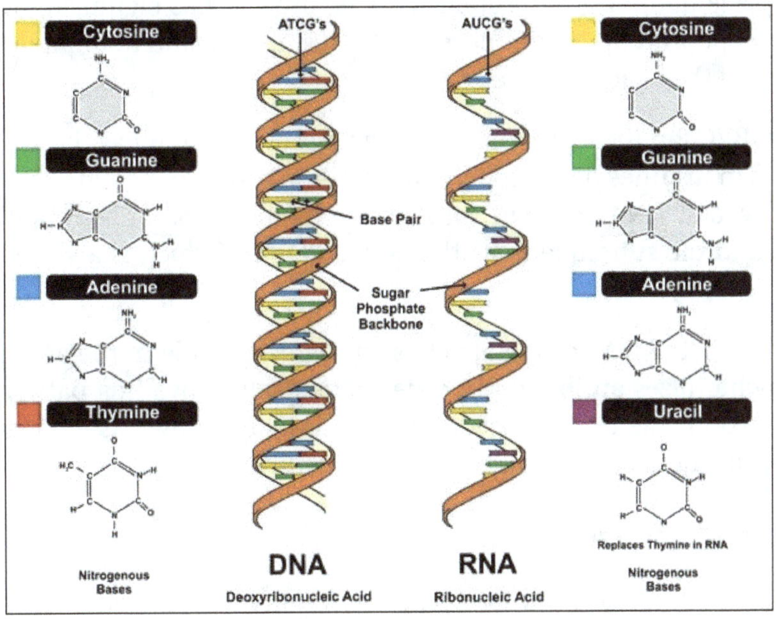

RNA editing occurs concurrently with transcription and splicing processes in the nucleus or mitochondria. It comes in a myriad of different natures, with all of the diverse editing types distributed intermittently across the phylogenetic spectrum.

A-to-I RNA Editing

RNA editing by adenosine-to-inosine (A-to-I) base modification successfully generates RNA and protein diversity in higher eukaryotes, selectively reshaping coding and non-coding sequences in nuclear transcripts. The enzymes responsible for this type of editing are known as ADARs (adenosine deaminases acting on RNA), which are pervasive in mammals.

Chemical conversion of adenosine to inosine occurs through removal of an amino group. Prior to editing adenosine hydrogen bonds with thymine, whereas after editing inosine hydrogen bonds with cytosine. A-to-I RNA editing often results in codon changes that alters the function of a chosen protein.

Most of ADAR editing events take place in non-coding regions of the human genome, and current estimates predict that there are more than 40 thousand editing sites in the human transcriptome. Messenger RNAs edited within coding regions at specific adenosines are only known to occur for approximately 30 human genes. Therefore the transcript diversity generated by RNA editing suitably affects regulatory elements in the messenger RNA.

C-to-U RNA Editing

A second important type of RNA editing in mammals also relies on base modification and it takes place in apoB (apoprotein B). This process has significant effects on the metabolism of lipoproteins, and it necessitates a single-strand template with clear-cut characteristics in the vicinity of the edited base.

Recently, APOBEC family of enzymes has been discovered that has the ability to deaminate cytidines to uridines on both RNA and DNA in order to mediate this type of RNA editing. This type of RNA editing results in the modification of a glutamine codon to a stop codon and the subsequent synthesis of APOB48 (which is a shorter isoform of APOB100).

APOBEC and C-to-U RNA editing also has an important role in the adaptive immune response, which utilizes antibodies in order to neutralize invading pathogens. Editing of antibody-encoding genes is carried out by an enzyme known as AID (activation-induced cytidine deaminase), also from APOBEC family.

RNA-editing Databases

RNA editing sites are already mapped on the reference human genome, and databases that supply all the available and published data linked to RNA editing can be found online. It is even possible to acquire RNA editing information for a specific RNA molecule or gene using matched accession numbers in RNA editing libraries.

Such endeavors already resulted in successful collaborations. For example, Thermo Fisher Scientific, which is one of the leading instrument and reagent providers, partnered with AstraZeneca and provided them with RNA-guide libraries that target individual gene families and human genes in order identify new biological targets.

Thermo Fisher Scientific is also trying to hasten the access to forefront genome-editing applications by employing RNA editing and other genome-modifying techniques.

Enabling more pertinent and applicable disease models will improve identification and editing of the target, which can in turn be translated to therapeutics.

AMS Biotechnology or AMSBIO, a company founded in 1987, is now one of the leaders in the field of genome editing, and also provides tissue nucleic acid and protein products for that purpose. Other commercial companies are increasingly joining the quest for RNA-based therapeutics and applications.

References

- Structure-and-function-of-rna, microbiology: courses.lumenlearning.com, Retrieved 31 March, 2020
- Quaresma AJ, Sievert R, Nickerson JA (April 2013). "Regulation of mrna export by the PI3 kinase/AKT signal transduction pathway". Molecular Biology of the Cell. 24 (8): 1208–21. Doi:10.1091/mbc.E12-06-0450. PMC 3623641. PMID 23427269
- Rna-properties-structure-types-and-functions: microbenotes.com, Retrieved 14 July, 2020
- Stryer L, Berg JM, Tymoczko JL (2002). Biochemistry (5th ed.). San Francisco: W.H. Freeman. ISBN 978-0-7167-4955-4
- Keiler KC (2008). "Biology of trans-translation". Annual Review of Microbiology. 62: 133–51. Doi:10.1146/annurev.micro.62.081307.162948. PMID 18557701
- Transcriptome-Fact-Sheet, fact-sheets, about-genomics: genome.gov, Retrieved 17 May, 2020
- Edmonds, M (2002). A history of poly A sequences: from formation to factors to function. Progress in Nucleic Acid Research and Molecular Biology. 71. Pp. 285–389. Doi:10.1016/S0079-6603(02)71046-5. ISBN 978-0-12-540071-8. PMID 12102557
- Types-of-RNA-Editing, life-sciences: news-medical.net, Retrieved 19 April, 2020
- Lodish, Havery (2004). Molecular Cell Biology. New York, New York: W.H. Freeman and Company. P. 113. ISBN 978-0-7167-4366-8
- Mirna-introduction, learning-center, functional-genomics-and-rnai, life-science: sigmaaldrich.com, Retrieved 21 May, 2020

3
DNA Replication

DNA replication refers to the process of producing a copy of DNA itself during cell division. DNA-binding proteins are proteins having DNA-binding domains and general affinity for single- or double-stranded DNA. This chapter will briefly introduce all the significant aspects of DNA replication.

Protein Synthesis

DNA provides a “blueprint” for the cell structure and physiology. This refers to the fact that DNA contains the information necessary for the cell to build one very important type of molecule: the protein. Most structural components of the cell are made up, at least in part, by proteins and virtually all the functions that a cell carries out are completed with the help of proteins. One of the most important classes of proteins is enzymes, which help speed up necessary biochemical reactions that take place inside the cell. Some of these critical biochemical reactions include building larger molecules from smaller components (such as occurs during DNA replication or synthesis of microtubules) and breaking down larger molecules into smaller components (such as when harvesting chemical energy from nutrient molecules). Whatever the cellular process may be, it is almost sure to involve proteins. Just as the cell’s genome describes its full complement of DNA, a cell’s proteome is its full complement of proteins. Protein synthesis begins with genes. A gene is a functional segment of DNA that provides the genetic information necessary to build a protein. Each particular gene provides the code necessary to construct a particular protein. Gene expression, which transforms the information coded in a gene to a final gene product, ultimately dictates the structure and function of a cell by determining which proteins are made.

The interpretation of genes works in the following way. Recall that proteins are polymers, or chains, of many amino acid building blocks. The sequence of bases in a gene (that is, its sequence of A, T, C, G nucleotides) translates to an amino acid sequence. A triplet is a section of three DNA bases in a row that codes for a specific amino acid. Similar to the way in which the three-letter code d-o-g signals the image of a dog, the three-letter DNA base code signals the use of a particular amino acid. For example, the DNA triplet CAC (cytosine, adenine, and cytosine) specifies the amino acid valine. Therefore, a gene, which is composed of multiple triplets in a unique sequence, provides the code to build an entire protein, with multiple amino acids in the proper sequence. The mechanism by which cells turn the DNA code into a protein product is a two-step process, with an RNA molecule as the intermediate.

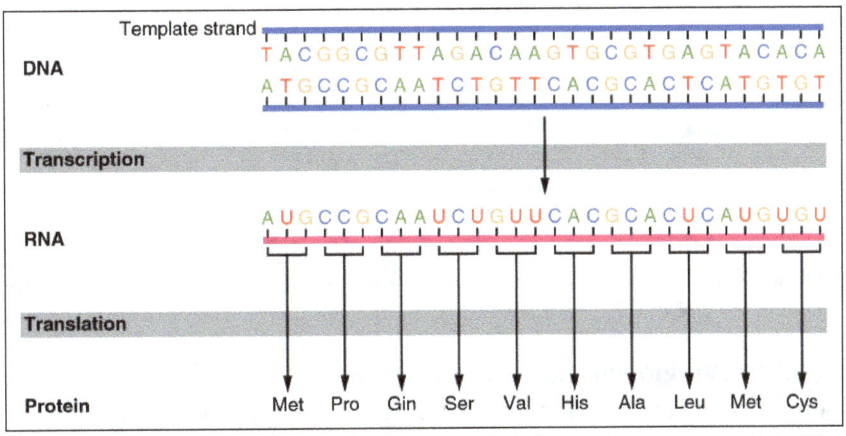

The Genetic Code. DNA holds all of the genetic information necessary to build a cell's proteins. The nucleotide sequence of a gene is ultimately translated into an amino acid sequence of the gene's corresponding protein.

From DNA to RNA: Transcription

DNA is housed within the nucleus, and protein synthesis takes place in the cytoplasm, thus there must be some sort of intermediate messenger that leaves the nucleus and manages protein synthesis. This intermediate messenger is messenger RNA (mRNA), a single-stranded nucleic acid that carries a copy of the genetic code for a single gene out of the nucleus and into the cytoplasm where it is used to produce proteins.

There are several different types of RNA, each having different functions in the cell. The structure of RNA is similar to DNA with a few small exceptions. For one thing, unlike DNA, most types of RNA, including mRNA, are single-stranded and contain no complementary strand. Second, the ribose sugar in RNA contains an additional oxygen atom compared with DNA. Finally, instead of the base thymine, RNA contains the base uracil. This means that adenine will always pair up with uracil during the protein synthesis process.

Gene expression begins with the process called transcription, which is the synthesis of a strand of mRNA that is complementary to the gene of interest. This process is called transcription because the mRNA is like a transcript, or copy, of the gene's DNA code. Transcription begins in a fashion somewhat like DNA replication, in that a region of DNA unwinds and the two strands separate, however, only that small portion of the DNA will be split apart. The triplets within the gene on this section of the DNA molecule are used as the template to transcribe the complementary strand of RNA. A codon is a three-base sequence of mRNA, so-called because they directly encode amino acids. Like DNA replication, there are three stages to transcription: initiation, elongation, and termination.

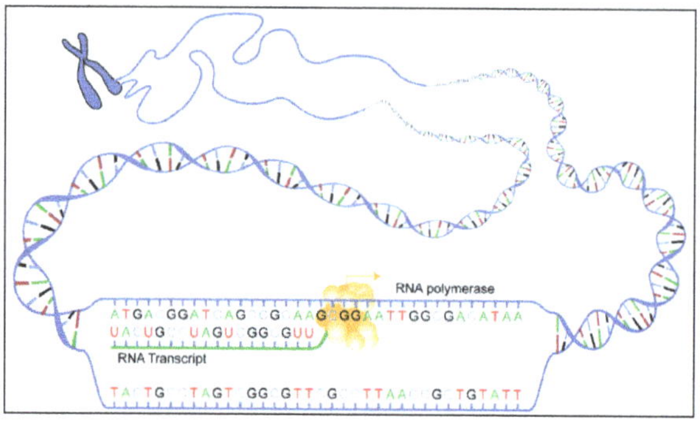

Transcription: from DNA to mRNA. In the first of the two stages of making protein from DNA, a gene on the DNA molecule is transcribed into a complementary mRNA molecule.

Stage 1 - Initiation: A region at the beginning of the gene called a promoter—a particular sequence of nucleotides—triggers the start of transcription.

Stage 2 - Elongation: Transcription starts when RNA polymerase unwinds the DNA segment. One strand, referred to as the coding strand, becomes the template with the genes to be coded. The polymerase then aligns the correct nucleic acid (A, C, G, or U) with its complementary base on the coding strand of DNA. RNA polymerase is an enzyme that adds new nucleotides to a growing strand of RNA. This process builds a strand of mRNA.

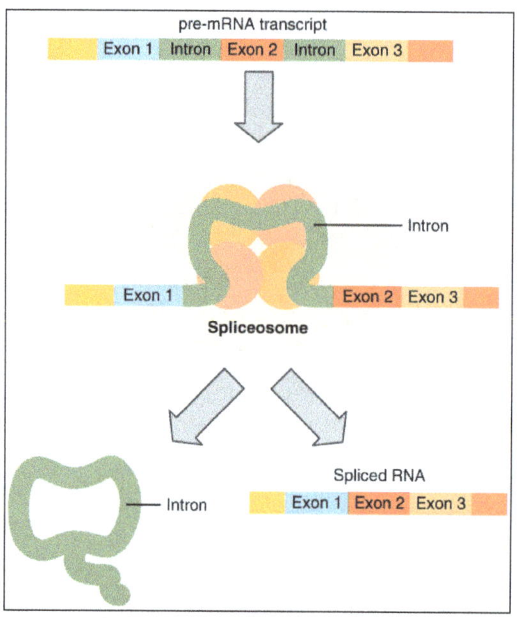

Splicing DNA. In the nucleus, a structure called a spliceosome cuts out introns (noncoding regions) within a pre-mRNA transcript and reconnects the exons.

Stage 3 - Termination: When the polymerase has reached the end of the gene, one of three specific triplets (UAA, UAG, or UGA) codes a "stop" signal, which triggers the enzymes to terminate transcription and release the mRNA transcript.

Before the mRNA molecule leaves the nucleus and proceeds to protein synthesis, it is modified in a number of ways. For this reason, it is often called a pre-mRNA at this stage. For example, your DNA, and thus complementary mRNA, contains long regions called non-coding regions that do not code for amino acids. Their function is still a mystery, but the process called splicing removes these non-coding regions from the pre-mRNA transcript. A spliceosome—a structure composed of various proteins and other molecules—attaches to the mRNA and "splices" or cuts out the non-coding regions. The removed segment of the transcript is called an intron. The remaining exons are pasted together. An exon is a segment of RNA that remains after splicing. Interestingly, some introns that are removed from mRNA are not always non-coding. When different coding regions of mRNA are spliced out, different variations of the protein will eventually result, with differences in structure and function. This process results in a much larger variety of possible proteins and protein functions. When the mRNA transcript is ready, it travels out of the nucleus and into the cytoplasm.

From RNA to Protein: Translation

Like translating a book from one language into another, the codons on a strand of mRNA must be translated into the amino acid alphabet of proteins. Translation is the process of synthesizing a chain of amino acids called a polypeptide. Translation requires two major aids: first, a "translator," the molecule that will conduct the translation, and second, a substrate on which the mRNA strand is translated into a new protein, like the translator's "desk." Both of these requirements are fulfilled by other types of RNA. The substrate on which translation takes place is the ribosome.

Remember that many of a cell's ribosomes are found associated with the rough ER, and carry out the synthesis of proteins destined for the Golgi apparatus. Ribosomal RNA (rRNA) is a type of RNA that, together with proteins, composes the structure of the ribosome. Ribosomes exist in the cytoplasm as two distinct components, a small and a large subunit. When an mRNA molecule is ready to be translated, the two subunits come together and attach to the mRNA. The ribosome provides a substrate for translation, bringing together and aligning the mRNA molecule with the molecular "translators" that must decipher its code.

The other major requirement for protein synthesis is the translator molecules that physically "read" the mRNA codons. Transfer RNA (tRNA) is a type of RNA that ferries the appropriate corresponding amino acids to the ribosome, and attaches each new amino acid to the last, building the polypeptide chain one-by-one. Thus tRNA transfers specific amino acids from the cytoplasm to a growing polypeptide. The tRNA molecules must be able to recognize the codons on mRNA and match them with the correct amino acid. The tRNA is modified for this function. On one end of its structure is a binding site for a specific amino acid. On the other end is a base sequence that matches the codon specifying its particular amino acid. This sequence of three bases on the tRNA molecule is called an anticodon. For example, a tRNA responsible for shuttling the amino acid glycine contains a binding site for glycine on one end. On the other end it contains an anticodon that complements the glycine codon (GGA is a codon for glycine, and so the tRNAs anticodon would read CCU). Equipped with its particular cargo and matching anticodon, a tRNA molecule can read its recognized mRNA codon and bring the corresponding amino acid to the growing chain.

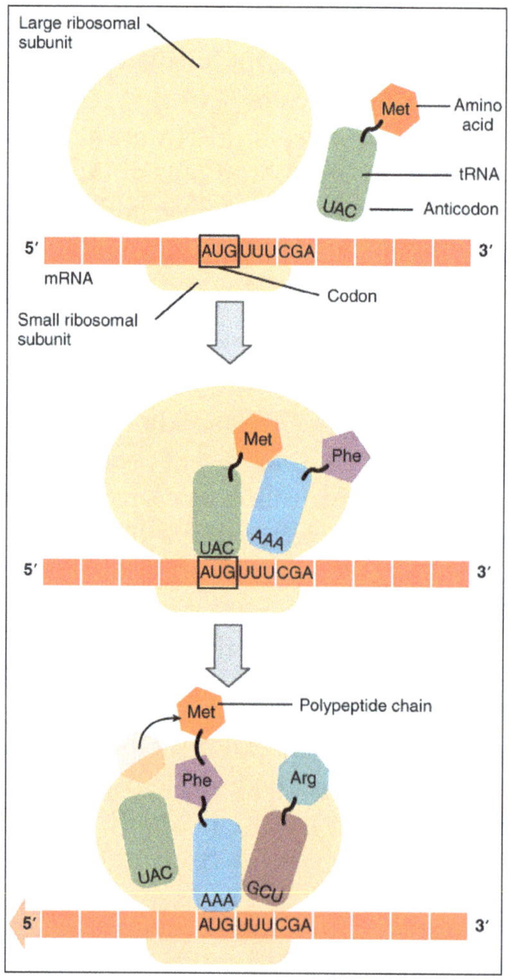

Translation from RNA to Protein. During translation, the mRNA transcript is "read" by a functional complex consisting of the ribosome and tRNA molecules. tRNAs bring the appropriate amino acids in sequence to the growing polypeptide chain by matching their anti-codons with codons on the mRNA strand.

Much like the processes of DNA replication and transcription, translation consists of three main stages: initiation, elongation, and termination. Initiation takes place with the binding of a ribosome to an mRNA transcript. The elongation stage involves the recognition of a tRNA anticodon with the next mRNA codon in the sequence. Once the anticodon and codon sequences are bound (remember, they are complementary base pairs), the tRNA presents its amino acid cargo and the growing polypeptide strand is attached to this next amino acid. This attachment takes place with the assistance of various enzymes and requires energy. The tRNA molecule then releases the mRNA strand, the mRNA strand shifts one codon over in the ribosome, and the next appropriate tRNA arrives with its matching anticodon. This process continues until the final codon on the mRNA is reached which provides a "stop" message that signals termination of translation and triggers the release of the complete, newly synthesized protein. Thus, a gene within the DNA molecule is transcribed into mRNA, which is then translated into a protein product.

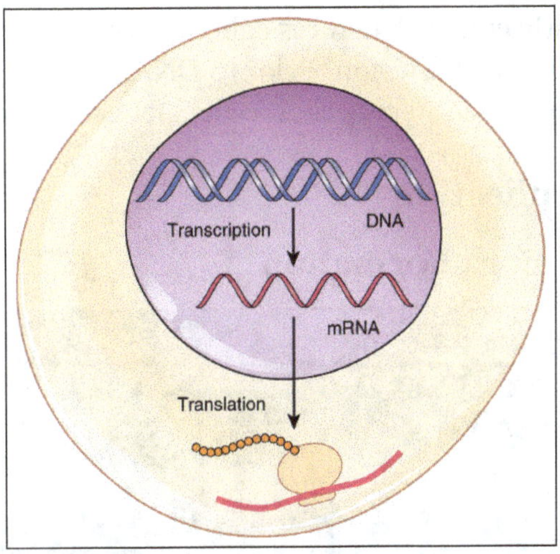

From DNA to Protein: Transcription through Translation. Transcription within the cell nucleus produces an mRNA molecule, which is modified and then sent into the cytoplasm for translation. The transcript is decoded into a protein with the help of a ribosome and tRNA molecules.

Commonly, an mRNA transcription will be translated simultaneously by several adjacent ribosomes. This increases the efficiency of protein synthesis. A single ribosome might translate an mRNA molecule in approximately one minute; so multiple ribosomes aboard a single transcript could produce multiple times the number of the same protein in the same minute. A polyribosome is a string of ribosomes translating a single mRNA strand.

DNA Replication

DNA is the genetic material that defines every cell. Before a cell duplicates and is divided into new daughter cells through either mitosis or meiosis, biomolecules and organelles must be copied to be distributed among the cells. DNA, found within the nucleus, must be replicated in order to ensure that each new cell receives the correct number of chromosomes. The process of DNA duplication is called DNA replication. Replication follows several steps that involve multiple proteins called replication enzymes and RNA. In eukaryotic cells, such as animal cells and plant cells, DNA replication occurs in the S phase of interphase during the cell cycle. The process of DNA replication is vital for cell growth, repair, and reproduction in organisms.

DNA Structure

DNA or deoxyribonucleic acid is a type of molecule known as a nucleic acid. It consists of a 5-carbon deoxyribose sugar, a phosphate, and a nitrogenous base. Double-stranded DNA consists of two spiral nucleic acid chains that are twisted into a double helix shape. This twisting allows DNA to be more compact. In order to fit within the nucleus, DNA is packed into tightly coiled structures called chromatin. Chromatin condenses to form chromosomes during cell division. Prior to DNA replication, the chromatin loosens giving cell replication machinery access to the DNA strands.

Preparation for Replication

Step 1: Replication Fork Formation

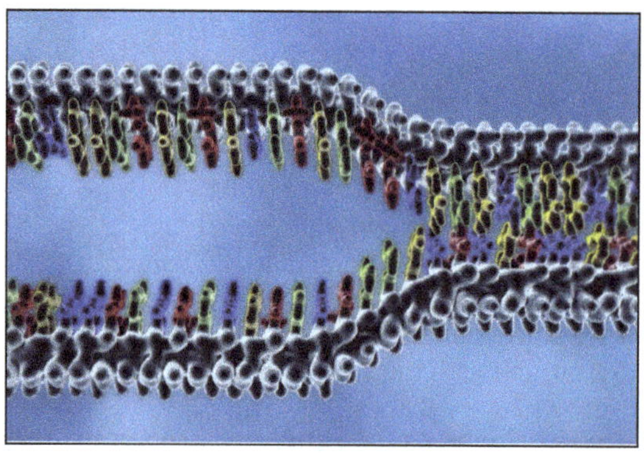

Before DNA can be replicated, the double stranded molecule must be "unzipped" into two single strands. DNA has four bases called adenine (A), thymine (T), cytosine (C) and guanine (G) that form pairs between the two strands. Adenine only pairs with thymine and cytosine only binds with guanine. In order to unwind DNA, these interactions between base pairs must be broken. This is performed by an enzyme known as DNA helicase. DNA helicase disrupts the hydrogen bonding between base pairs to separate the strands into a Y shape known as the replication fork. This area will be the template for replication to begin.

DNA is directional in both strands, signified by a 5' and 3' end. This notation signifies which side group is attached the DNA backbone. The 5' end has a phosphate (P) group attached, while the 3' end has a hydroxyl (OH) group attached. This directionality is important for replication as it only progresses in the 5' to 3' direction. However, the replication fork is bi-directional; one strand is oriented in the 3' to 5' direction (leading strand) while the other is oriented 5' to 3' (lagging strand). The two sides are therefore replicated with two different processes to accommodate the directional difference.

Replication Begins

Step 2: Primer Binding

The leading strand is the simplest to replicate. Once the DNA strands have been separated, a short piece of RNA called a primer binds to the 3' end of the strand. The primer always binds as the starting point for replication. Primers are generated by the enzyme DNA primase.

DNA Replication: Elongation

Step 3: Elongation

Enzymes known as DNA polymerases are responsible creating the new strand by a process called elongation. There are five different known types of DNA polymerases in bacteria and human cells. In bacteria such as E. coli, polymerase III is the main replication enzyme, while polymerase I, II, IV and V are responsible for error checking and repair. DNA polymerase III binds to the strand at the site of the primer and begins adding new base pairs complementary to the strand during replication. In eukaryotic cells, polymerases alpha, delta, and epsilon are the primary polymerases involved in DNA replication. Because replication proceeds in the 5' to 3' direction on the leading strand, the newly formed strand is continuous.

DNA polymerases (blue) attach themselves to the DNA and elongate the new strands by adding nucleotide bases.

The lagging strand begins replication by binding with multiple primers. Each primer is only several bases apart. DNA polymerase then adds pieces of DNA, called Okazaki fragments, to the strand between primers. This process of replication is discontinuous as the newly created fragments are disjointed.

Step 4: Termination

Once both the continuous and discontinuous strands are formed, an enzyme called exonuclease removes all RNA primers from the original strands. These primers are then replaced with appropriate bases. Another exonuclease "proofreads" the newly formed DNA to check, remove and replace any errors. Another enzyme called DNA ligase joins Okazaki fragments together forming a single unified strand. The ends of the linear DNA present a problem as DNA polymerase can only add nucleotides in the 5′ to 3′ direction. The ends of the parent strands consist of repeated DNA sequences called telomeres. Telomeres act as protective caps at the end of chromosomes to prevent nearby chromosomes from fusing. A special type of DNA polymerase enzyme called telomerase catalyzes the synthesis of telomere sequences at the ends of the DNA. Once completed, the parent strand and its complementary DNA strand coils into the familiar double helix shape. In the end, replication produces two DNA molecules, each with one strand from the parent molecule and one new strand.

Replication Enzymes

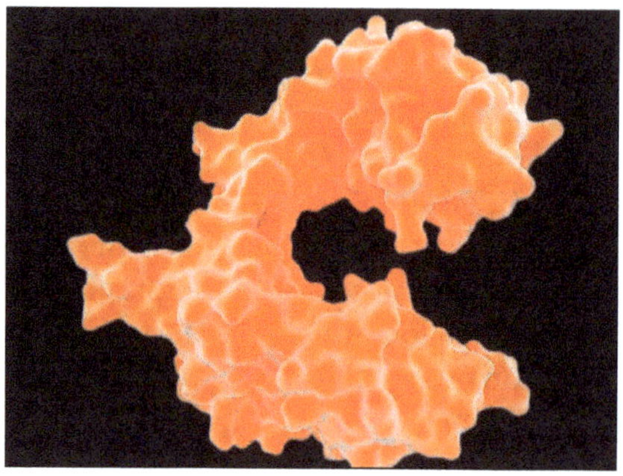

DNA polymerase molecule.

DNA replication would not occur without enzymes that catalyze various steps in the process. Enzymes that participate in the eukaryotic DNA replication process include:

- DNA helicase: Unwinds and separates double stranded DNA as it moves along the DNA. It forms the replication fork by breaking hydrogen bonds between nucleotide pairs in DNA.
- DNA primase: A type of RNA polymerase that generates RNA primers. Primers are short RNA molecules that act as templates for the starting point of DNA replication.
- DNA polymerases: Synthesize new DNA molecules by adding nucleotides to leading and lagging DNA strands.
- Topoisomerase or DNA Gyrase: Unwinds and rewinds DNA strands to prevent the DNA from becoming tangled or supercoiled.
- Exonucleases: Group of enzymes that remove nucleotide bases from the end of a DNA chain.

- DNA ligase: Joins DNA fragments together by forming phosphodiester bonds between nucleotides.

DNA Replication

DNA replication is the production of identical DNA helices from a single double-stranded DNA molecule. Each molecule consists of a strand from the original molecule and a newly formed strand. Prior to replication, the DNA uncoils and strands separate. A replication fork is formed which serves as a template for replication. Primers bind to the DNA and DNA polymerases add new nucleotide sequences in the 5′ to 3′ direction.

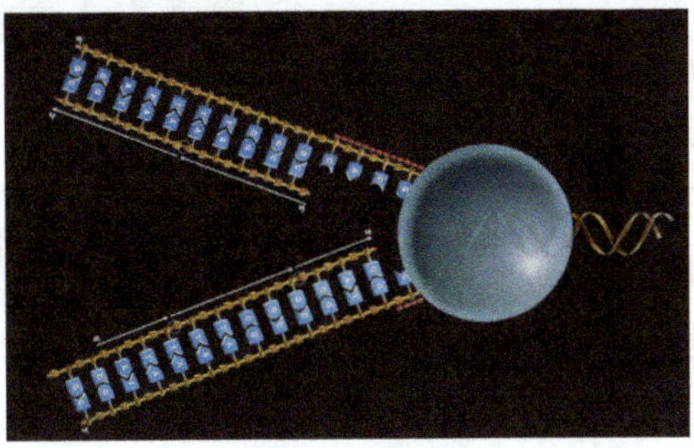

Replication of DNA.

This addition is continuous in the leading strand and fragmented in the lagging strand. Once elongation of the DNA strands is complete, the strands are checked for errors, repairs are made, and telomere sequences are added to the ends of the DNA.

DNA-binding Protein

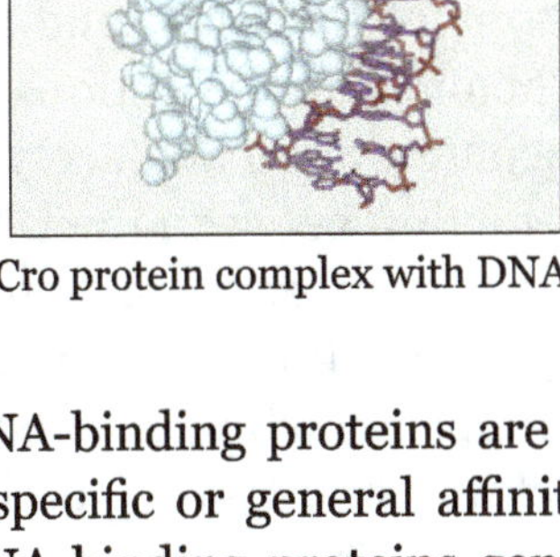

Cro protein complex with DNA.

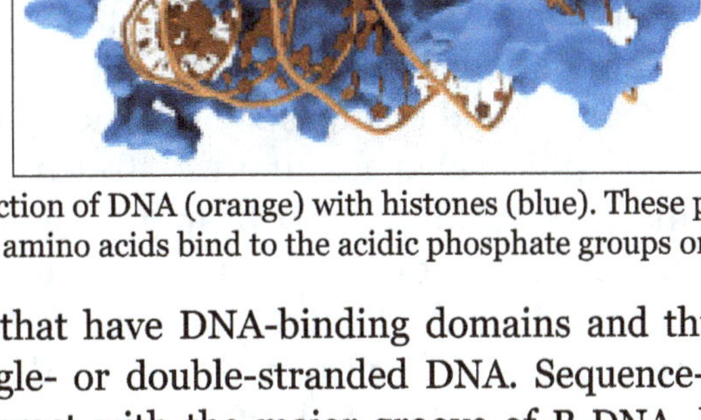

Interaction of DNA (orange) with histones (blue). These proteins' basic amino acids bind to the acidic phosphate groups on DNA.

DNA-binding proteins are proteins that have DNA-binding domains and thus have a specific or general affinity for single- or double-stranded DNA. Sequence-specific DNA-binding proteins generally interact with the major groove of B-DNA, because it exposes more functional groups that identify a base pair. However, there are some known minor groove DNA-binding ligands such as netropsin, distamycin, Hoechst 33258, pentamidine, DAPI and others.

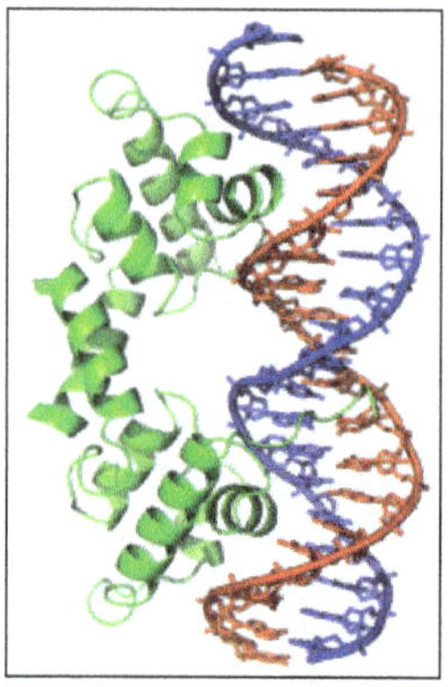

The lambda repressor helix-turn-helix transcription factor bound to its DNA target.

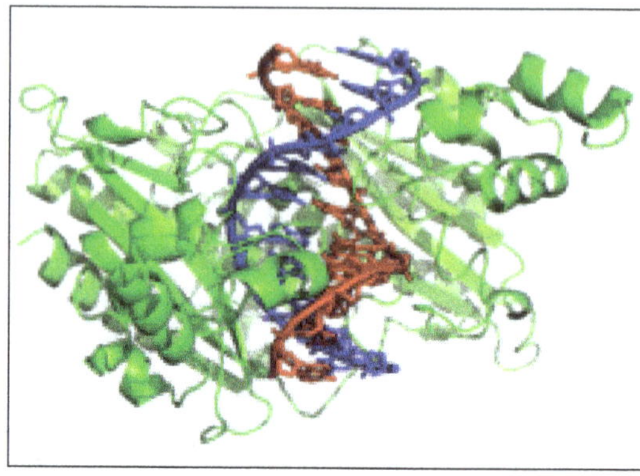

The restriction enzyme EcoRV (green) in a complex with its substrate DNA.

Examples: DNA-binding proteins include transcription factors which modulate the process of transcription, various polymerases, nucleases which cleave DNA molecules, and histones which are involved in chromosome packaging and transcription in the cell nucleus. DNA-binding proteins can incorporate such domains as the zinc finger, the helix-turn-helix, and the leucine zipper (among many others) that facilitate binding to nucleic acid. There are also more unusual examples such as transcription activator like effectors.

Non-specific DNA-protein Interactions

Structural proteins that bind DNA are well-understood examples of non-specific DNA-protein interactions. Within chromosomes, DNA is held in complexes with structural proteins. These proteins organize the DNA into a compact structure called chromatin. In eukaryotes, this structure involves DNA binding to a complex of small basic proteins called histones. In prokaryotes, multiple types of proteins are involved. The histones form a disk-shaped complex called a nucleosome, which contains two complete turns of double-stranded DNA wrapped around its surface. These non-specific interactions are formed through basic residues in the histones making ionic bonds to the acidic sugar-phosphate backbone of the DNA, and are therefore largely independent of the base sequence. Chemical modifications of these basic amino acid residues include methylation, phosphorylation and acetylation. These chemical changes alter the strength of the interaction between the DNA and the histones, making the DNA more or less accessible to transcription factors and changing the rate of transcription. Other non-specific DNA-binding proteins in chromatin include the high-mobility group (HMG) proteins, which bind to bent or distorted DNA. Biophysical studies show that these architectural HMG proteins bind, bend and loop DNA to perform its biological functions. These proteins are important in bending arrays of nucleosomes and arranging them into the larger structures that form chromosomes.

Binding to Specific DNA Sequences

In contrast, other proteins have evolved to bind to specific DNA sequences. The most intensively studied of these are the various transcription factors, which are proteins that regulate transcription. Each transcription factor binds to one specific set of DNA sequences and activates or inhibits the transcription of genes that have these sequences near their promoters. The transcription factors do this in two ways. Firstly, they can bind the RNA polymerase responsible for transcription, either directly or through other mediator proteins; this locates the polymerase at the promoter and allows it to begin transcription. Alternatively, transcription factors can bind enzymes that modify the histones at the promoter. This alters the accessibility of the DNA template to the polymerase.

These DNA targets can occur throughout an organism's genome. Thus, changes in the activity of one type of transcription factor can affect thousands of genes. Thus, these proteins are often the targets of the signal transduction processes that control responses to environmental changes or cellular differentiation and development. The specificity of these transcription factors' interactions with DNA come from the proteins making multiple contacts to the edges of the DNA bases, allowing them to *read* the DNA sequence. Most of these base-interactions are made in the major groove, where the bases are most accessible. Mathematical descriptions of protein-DNA binding taking into account sequence-specificity, and competitive and cooperative binding of proteins of different types are usually performed with the help of the lattice models. Computational methods to identify the DNA binding sequence specificity have been proposed to make a good use of the abundant sequence data in the post-genomic era.

Protein–DNA Interactions

Protein–DNA interactions occur when a protein binds a molecule of DNA, often to regulate the biological function of DNA, usually the expression of a gene. Among the proteins that bind to DNA are transcription factors that activate or repress gene expression by binding to DNA motifs and histones that form part of the structure of DNA and bind to it less specifically. Also proteins that repair DNA such as uracil-DNA glycosylase interact closely with it.

In general, proteins bind to DNA in the major groove; however, there are exceptions. Protein–DNA interaction are of mainly two types, either specific interaction, or non-specific interaction. Recent single-molecule experiments showed that DNA binding proteins undergo of rapid rebinding in order to bind in correct orientation for recognizing the target site.

Design

Designing DNA-binding proteins that have a specified DNA-binding site has been an important goal for biotechnology. Zinc finger proteins have been designed to bind to specific DNA sequences and this is the basis of zinc finger nucleases. Recently transcription activator-like effector nucleases (TALENs) have been created which are based on natural proteins secreted by *Xanthomonas* bacteria via their type III secretion system when they infect various plant species.

Detection Methods

There are many *in vitro* and *in vivo* techniques which are useful in detecting DNA-Protein Interactions. The following lists some methods currently in use: Electrophoretic mobility shift assay is a widespread technique to identify protein–DNA interactions. DNase footprinting assay can be used to identify the specific site of binding of a protein to DNA. Chromatin immunoprecipitation is used to identify the sequence of the DNA fragments which bind to a known transcription factor. This technique when combined with high throughput sequencing is known as ChIP-Seq and when combined with microarrays it is known as ChIP-chip. Yeast one-hybrid System (Y1H) is used to identify which protein binds to a particular DNA fragment. Bacterial one-hybrid system (B1H) is used to identify which protein binds to a particular DNA fragment. Structure determination using X-ray crystallography has been used to give a highly detailed atomic view of protein–DNA interactions.

Manipulating the Interactions

The protein–DNA interactions can be modulated using stimuli like ionic strength of the buffer, macromolecular crowding, temperature, pH and electric field. This can lead to reversible dissociation/association of the protein–DNA complex.

Single-stranded Binding Protein

Single-stranded binding proteins (SSBs) are a class of proteins that have been identified in both viruses and organisms from bacteria to humans.

Viral SSB

Although the overall picture of *human cytomegalovirus* (HHV-5) DNA synthesis appears typical of the herpesviruses, some novel features are emerging.

Structure

In ICP8, the herpes simplex virus (HSV-1) single-strand DNA-binding protein (ssDNA-binding protein (SSB)), the head consists of the eight alpha helices. The front side of the neck region consists of a five-stranded beta-sheet and two alpha helices, whereas the back side is a three-stranded beta-sheet The shoulder part of the N-terminal domain contains an alpha-helical and beta-sheet region. The herpes simplex virus (HSV-1) SSB, ICP8, is a nuclear protein that, along other replication proteins is required for viral DNA replication during lytic infection.

Mechanism

Six herpes virus-group-common genes encode proteins that likely constitute the replication fork machinery, including a two-subunit DNA polymerase, a Helicase-primase complex and a single-stranded DNA-binding protein. The *human herpesvirus 1* (HHV-1) single-strand DNA-binding protein ICP8 is a 128kDa zinc metalloprotein. Photoaffinity labeling has shown that the region encompassing amino acid residues 368-902 contains the single-strand DNA-binding site of ICP8. The HHHV-1 UL5, UL8, and UL52 genes encode an essential heterotrimeric DNA helicase-primase that is responsible for concomitant DNA unwinding and primer synthesis at the viral DNA replication fork. ICP8 may stimulate DNA unwinding and enable bypass of cisplatin damaged DNA by recruiting the helicase-primase to the DNA.

Bacterial SSB

SSB protein domains in bacteria are important maintaining DNA metabolism, more specifically DNA replication, repair and recombination. It has a structure of three beta-strands to a single six-stranded beta-sheet to form a dimer.

Eukaryotic Replication Protein A

Replication protein A is the functional equivalent of SSB in the nucleus of eukaryotic cells, though there is no sequence homology.

Eukaryotic Mitochondrial SSB

The mitochondria of eukaryotic cells contain their own single stranded DNA binding protein. Human mitochondrial SSB (mtSSB) binds to single-stranded mitochondrial DNA as a tetramer and has sequence similarity to bacterial SSB. Human mtSSB is encoded by the SSBP1 gene. In yeast, it is encoded by the RIM1 gene.

Role in Genome Repair and Anti-aging

Recently we are finding that 1. it helps protect the genome, 2. is vital for stem cells and 3. involved with maintaining telomere length.

Histone

In biology, histones are highly alkaline proteins found in eukaryotic cell nuclei that package and order the DNA into structural units called nucleosomes. They are the chief protein components of chromatin, acting as spools around which DNA winds, and playing a role in gene regulation. Without histones, the unwound DNA in chromosomes would be very long (a length to width ratio of more than 10 million to 1 in human DNA). For example, each human diploid cell (containing 23 pairs of chromosomes) has about 1.8 meters of DNA; wound on the histones, the diploid cell has about 90 micrometers (0.09 mm) of chromatin. When the diploid cells are duplicated and condensed during mitosis, the result is about 120 micrometers of chromosomes.

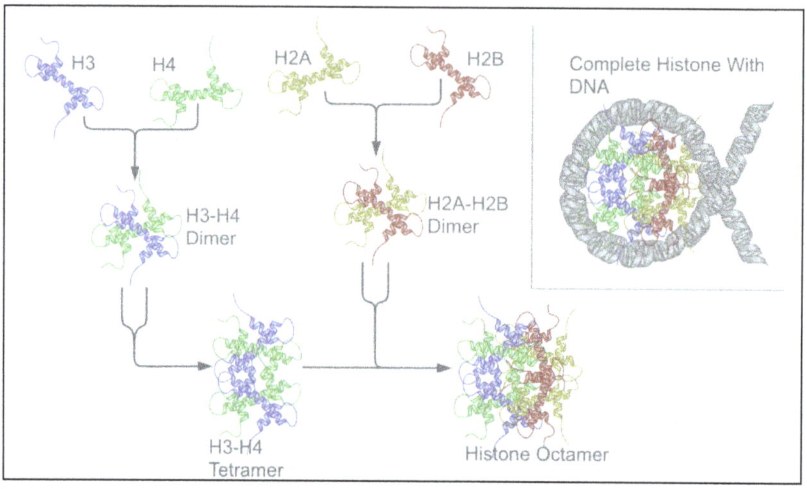

Schematic representation of the assembly of the core histones into the nucleosome.

Classes and Histone Variants

Five major families of histones exist: H1/H5, H2A, H2B, H3, and H4. Histones H2A, H2B, H3 and H4 are known as the core histones, while histones H1/H5 are known as the linker histones.

The core histones all exist as dimers, which are similar in that they all possess the histone fold domain: three alpha helices linked by two loops. It is this helical structure that allows for interaction between distinct dimers, particularly in a head-tail fashion (also called the handshake motif). The resulting four distinct dimers then come together to form one octameric nucleosome core, approximately 63 Angstroms in diameter (a solenoid (DNA)-like particle). Around 146 base pairs (bp) of DNA wrap around this core particle 1.65 times in a left-handed super-helical turn to give a particle of around 100 Angstroms across. The linker histone H1 binds the nucleosome at the entry and exit sites of the DNA, thus locking the DNA into place and allowing the formation of higher order structure. The most basic such formation is the 10 nm fiber or beads on a string conformation. This involves the wrapping of DNA around nucleosomes with approximately 50 base pairs of DNA separating each pair of nucleosomes (also referred to as linker DNA). Higher-order structures include the 30 nm fiber (forming an irregular zigzag) and 100 nm fiber, these being the structures found in normal cells. During mitosis and meiosis, the condensed chromosomes are assembled through interactions between nucleosomes and other regulatory proteins.

Histones are subdivided into canonical replication-dependent histones that are expressed during the S-phase of cell cycle and replication-independent histone variants, expressed during the whole cell cycle. In animals, genes encoding canonical histones are typically clustered along the chromosome, lack introns and use a stem loop structure at the 3' end instead of a polyA tail. Genes encoding histone variants are usually not clustered, have introns and their mRNAs are regulated with polyA tails. Complex multicellular organisms typically have a higher number of histone variants providing a variety of different functions. Recent data are accumulating about the roles of diverse histone variants highlighting the functional links between variants and the delicate regulation of organism development. Histone variants from different organisms, their classification and variant specific features can be found in "HistoneDB 2.0 - Variants" database.

The following is a list of human histone proteins:

Super family	Family	Subfamily	Members
Linker	H1	H1F	H1F0, H1FNT, H1FOO, H1FX
		H1H1	HIST1H1A, HIST1H1B, HIST1H1C, HIST1H1D, HIST1H1E, HIST1H1T
Core	H2A	H2AF	H2AFB1, H2AFB2, H2AFB3, H2AFJ, H2AFV, H2AFX, H2AFY, H2AFY2, H2AFZ
		H2A1	HIST1H2AA, HIST1H2AB, HIST1H2AC, HIST1H2AD, HIST1H2AE, HIST1H2AG, HIST1H2AI, HIST1H2AJ, HIST1H2AK, HIST1H2AL, HIST1H2AM
		H2A2	HIST2H2AA3, HIST2H2AC
	H2B	H2BF	H2BFM, H2BFS, H2BFWT
		H2B1	HIST1H2BA, HIST1H2BB, HIST1H2BC, HIST1H2BD, HIST1H2BE, HIST1H2BF, HIST1H2BG, HIST1H2BH, HIST1H2BI, HIST1H2BJ, HIST1H2BK, HIST1H2BL, HIST1H2BM, HIST1H2BN, HIST1H2BO
		H2B2	HIST2H2BE
	H3	H3A1	HIST1H3A, HIST1H3B, HIST1H3C, HIST1H3D, HIST1H3E, HIST1H3F, HIST1H3G, HIST1H3H, HIST1H3I, HIST1H3J
		H3A2	HIST2H3C
		H3A3	HIST3H3
	H4	H41	HIST1H4A, HIST1H4B, HIST1H4C, HIST1H4D, HIST1H4E, HIST1H4F, HIST1H4G, HIST1H4H, HIST1H4I, HIST1H4J, HIST1H4K, HIST1H4L
		H44	HIST4H4

Structure

The nucleosome core is formed of two H2A-H2B dimers and a H3-H4 tetramer, forming two nearly symmetrical halves by tertiary structure (C2 symmetry; one macromolecule is the mirror image of the other). The H2A-H2B dimers and H3-H4 tetramer also show pseudodyad symmetry. The 4 'core' histones (H2A, H2B, H3 and H4) are relatively similar in structure and are highly conserved through evolution, all featuring a 'helix turn helix turn helix' motif (DNA-binding protein motif that recognize specific DNA sequence). They also share the feature of long 'tails' on one end of the amino acid structure - this being the location of post-translational modification.

Archaeal histone only contains a H3-H4 like dimeric structure made out of the same protein. Such dimeric structures can stack into a tall superhelix ("supernucleosome") onto which DNA coils in a manner similar to nucleosome spools. Only some archaeal histones have tails.

It has been proposed that histone proteins are evolutionarily related to the helical part of the extended AAA+ ATPase domain, the C-domain, and to the N-terminal substrate recognition domain of Clp/Hsp100 proteins. Despite the differences in their topology, these three folds share a homologous helix-strand-helix (HSH) motif.

Using an electron paramagnetic resonance spin-labeling technique, British researchers measured the distances between the spools around which eukaryotic cells wind their DNA. They determined the spacings range from 59 to 70 Å.

In all, histones make five types of interactions with DNA:

- Helix-dipoles form alpha-helixes in H2B, H3, and H4 cause a net positive charge to accumulate at the point of interaction with negatively charged phosphate groups on DNA.
- Hydrogen bonds between the DNA backbone and the amide group on the main chain of histone proteins.
- Nonpolar interactions between the histone and deoxyribose sugars on DNA.
- Salt bridges and hydrogen bonds between side chains of basic amino acids (especially lysine and arginine) and phosphate oxygens on DNA.
- Non-specific minor groove insertions of the H3 and H2B N-terminal tails into two minor grooves each on the DNA molecule.

The highly basic nature of histones, aside from facilitating DNA-histone interactions, contributes to their water solubility.

Histones are subject to post translational modification by enzymes primarily on their N-terminal tails, but also in their globular domains. Such modifications include methylation, citrullination, acetylation, phosphorylation, SUMOylation, ubiquitination, and ADP-ribosylation. This affects their function of gene regulation.

In general, genes that are active have less bound histone, while inactive genes are highly associated with histones during interphase. It also appears that the structure of histones has been evolutionarily conserved, as any deleterious mutations would be severely maladaptive. All histones have a highly positively charged N-terminus with many lysine and arginine residues.

Conservation across Species

Histones are found in the nuclei of eukaryotic cells, and in certain Archaea, namely Proteoarchaea and Euryarchaea, but not in bacteria. The unicellular algae known as dinoflagellates were previously thought to be the only eukaryotes that completely lack histones, however, later studies showed that their DNA still encodes histone genes. Unlike the core histones, lysine-rich linker histone (H1) proteins are found in bacteria, otherwise known as nucleoprotein HC1/HC2.

Archaeal histones may well resemble the evolutionary precursors to eukaryotic histones. Histone proteins are among the most highly conserved proteins in eukaryotes, emphasizing their important role in the biology of the nucleus. In contrast mature sperm cells largely use protamines to package their genomic DNA, most likely because this allows them to achieve an even higher packaging ratio.

There are some *variant* forms in some of the major classes. They share amino acid sequence homology and core structural similarity to a specific class of major histones but also have their own feature that is distinct from the major histones. These *minor histones* usually carry out specific functions of the chromatin metabolism. For example, histone H3-like CENPA is associated with only the centromere region of the chromosome. Histone H2A variant H2A.Z is associated with the promoters of actively transcribed genes and also involved in the prevention of the spread of silent heterochromatin. Furthermore, H2A.Z has roles in chromatin for genome stability. Another H2A variant H2A.X is phosphorylated at S139 in regions around double-strand breaks and marks the region undergoing DNA repair. Histone H3.3 is associated with the body of actively transcribed genes.

Evolutionary Origin

Histones are believed to have evolved from ribosomal proteins with which they share much in common, both being short and basic proteins.

Function

Compacting DNA Strands

Histones act as spools around which DNA winds. This enables the compaction necessary to fit the large genomes of eukaryotes inside cell nuclei: the compacted molecule is 40,000 times shorter than an unpacked molecule.

Chromatin Regulation

Histones undergo posttranslational modifications that alter their interaction with DNA and nuclear proteins. The H3 and H4 histones have long tails protruding from the nucleosome, which can be covalently modified at several places. Modifications of the tail include methylation, acetylation, phosphorylation, ubiquitination, SUMOylation, citrullination, and ADP-ribosylation. The core of the histones H2A and H2B can also be modified. Combinations of modifications are thought to constitute a code, the so-called "histone code". Histone modifications act in diverse biological processes such as gene regulation, DNA repair, chromosome condensation (mitosis) and spermatogenesis (meiosis).

The common nomenclature of histone modifications is:

- The name of the histone (e.g., H3).
- The single-letter amino acid abbreviation (e.g., K for Lysine) and the amino acid position in the protein.
- The type of modification (Me: methyl, P: phosphate, Ac: acetyl, Ub: ubiquitin).
- The number of modifications (only Me is known to occur in more than one copy per residue. 1, 2 or 3 is mono-, di- or tri-methylation).

So H3K4me1 denotes the monomethylation of the 4th residue (a lysine) from the start (i.e., the N-terminal) of the H3 protein.

Table: Examples of histone modifications in transcriptional regulation.

Type of modification	Histone								
	H3K4	H3K9	H3K14	H3K27	H3K79	H3K36	H4K20	H2BK5	H2BK20
mono-methylation	activation	activation		activation	activation		activation	activation	
di-methylation		repression		repression	activation				
tri-methylation	activation	repression		repression	activation, repression	activation		repression	
acetylation	activation	activation	activation	activation					activation

Functions of Histone Modifications

A huge catalogue of histone modifications have been described, but a functional understanding of most is still lacking. Collectively, it is thought that histone modifications may underlie a histone code, whereby combinations of histone modifications have specific meanings. However, most functional data concerns individual prominent histone modifications that are biochemically amenable to detailed study.

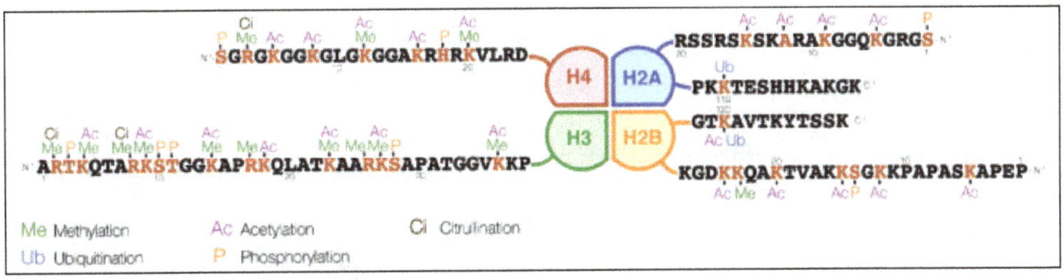

Chemistry of Histone Modifications

Lysine Methylation

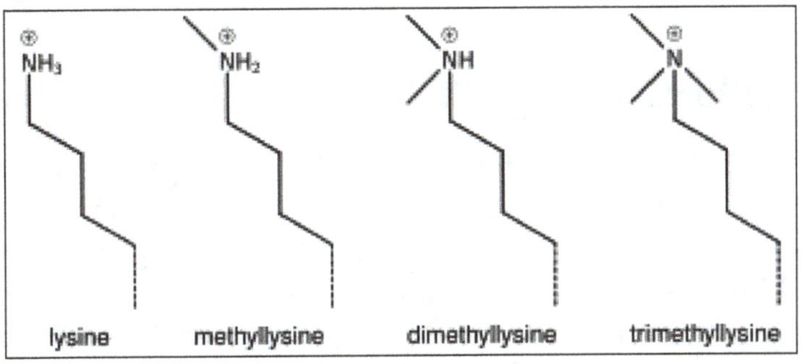

The addition of one, two, or many methyl groups to lysine has little effect on the chemistry of the histone; methylation leaves the charge of the lysine intact and adds a minimal number of atoms so steric interactions are mostly unaffected. However, proteins containing Tudor, chromo or PHD domains, amongst others, can recognise lysine methylation with exquisite sensitivity and differentiate mono, di and tri-methyl lysine, to the extent that, for some lysines (e.g.: H4K20) mono, di and tri-methylation appear to have different meanings. Because of this, lysine methylation tends to be a very informative mark and dominates the known histone modification functions.

Glutamine Serotonylation

Recently it has been shown, that the addition of a serotonin group to the position 5 glutamine of H3, happens in serotonergic cells such as neurons. This is part of the differentiation of the serotonergic cells. This post-translational modification happens in conjunction with the H3K4me3 modification. The serotonylation potentiates the binding of the general transcription factor TFIID to the TATA box.

Arginine Methylation

What was said above of the chemistry of lysine methylation also applies to arginine methylation, and some protein domains—e.g., Tudor domains—can be specific for methyl arginine instead of methyl lysine. Arginine is known to be mono- or di-methylated, and methylation can be symmetric or asymmetric, potentially with different meanings.

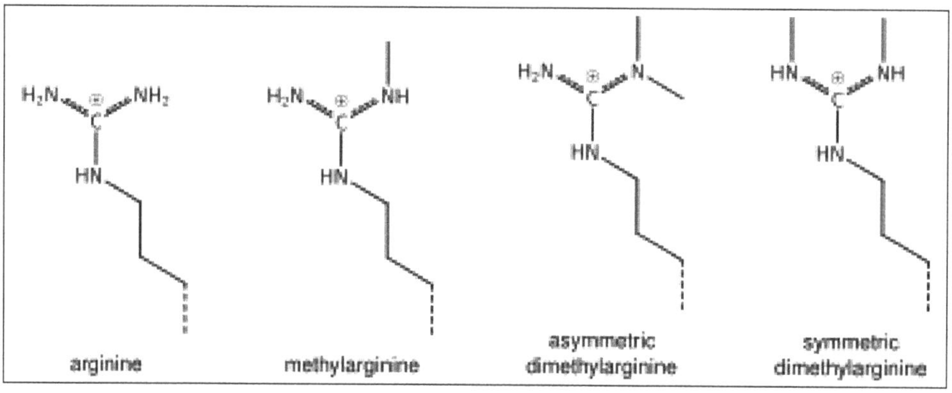

Arginine Citrullination

Enzymes called peptidylarginine deiminases (PADs) hydrolyze the imine group of arginines and attach a keto group, so that there is one less positive charge on the amino acid residue. This process has been involved in the activation of gene expression by making the modified histones less tightly bound to DNA and thus making the chromatin more accessible. PADs can also produce the opposite effect by removing or inhibiting mono-methylation of arginine residues on histones and thus antagonizing the positive effect arginine methylation has on transcriptional activity.

Lysine Acetylation

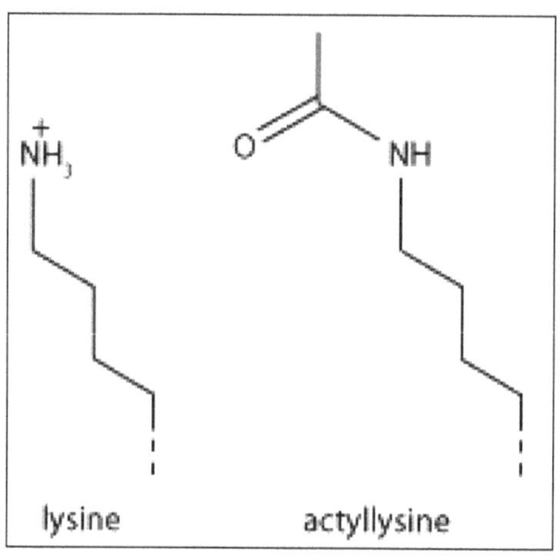

Addition of an acetyl group has a major chemical effect on lysine as it neutralises the positive charge. This reduces electrostatic attraction between the histone and the negatively charged DNA backbone, loosening the chromatin structure; highly acetylated histones form more accessible chromatin and tend to be associated with active transcription. Lysine acetylation appears to be less precise in meaning than methylation, in that histone acetyltransferases tend to act on more than one lysine; presumably this reflects the need to alter multiple lysines to have a significant effect on chromatin structure. The modification includes H3K27ac.

Serine/Threonine/Tyrosine Phosphorylation

Addition of a negatively charged phosphate group can lead to major changes in protein structure, leading to the well-characterised role of phosphorylation in controlling protein function. It is not clear what structural implications histone phosphorylation has, but histone phosphorylation has clear functions as a post-translational modification, and binding domains such as BRCT have been characterised.

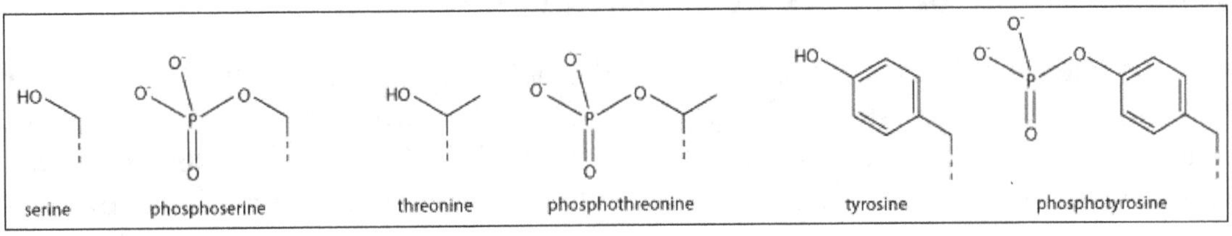

Functions in Transcription

Most well-studied histone modifications are involved in control of transcription.

Actively Transcribed Genes

Two histone modifications are particularly associated with active transcription:

Trimethylation of H3 Lysine 4 (H3K4me3)

This trimethylation occurs at the promoter of active genes and is performed by the COMPASS complex. Despite the conservation of this complex and histone modification from yeast to mammals, it is not entirely clear what role this modification plays. However, it is an excellent mark of active promoters and the level of this histone modification at a gene's promoter is broadly correlated with transcriptional activity of the gene. The formation of this mark is tied to transcription in a rather convoluted manner: early in transcription of a gene, RNA polymerase II undergoes a switch from initiating' to 'elongating', marked by a change in the phosphorylation states of the RNA polymerase II C terminal domain (CTD). The same enzyme that phosphorylates the CTD also phosphorylates the Rad6 complex, which in turn adds a ubiquitin mark to H2B K123 (K120 in mammals). H2BK123Ub occurs throughout transcribed regions, but this mark is required for COMPASS to trimethylate H3K4 at promoters.

Trimethylation of H3 Lysine 36 (H3K36me3)

This trimethylation occurs in the body of active genes and is deposited by the methyltransferase Set2. This protein associates with elongating RNA polymerase II, and H3K36Me3 is indicative of actively transcribed genes. H3K36Me3 is recognised by the Rpd3 histone deacetylase complex, which removes acetyl modifications from surrounding histones, increasing chromatin compaction and repressing spurious transcription. Increased chromatin compaction prevents transcription factors from accessing DNA, and reduces the likelihood of new transcription events being initiated within the body of the gene. This process therefore helps ensure that transcription is not interrupted.

Repressed Genes

Three histone modifications are particularly associated with repressed genes:

Trimethylation of H3 Lysine 27 (H3K27me3)

This histone modification is depositied by the polycomb complex PRC2. It is a clear marker of gene repression, and is likely bound by other proteins to exert a repressive function. Another polycomb complex, PRC1, can bind H3K27me3 and adds the histone modification H2AK119Ub which aids chromatin compaction. Based on this data it appears that PRC1 is recruited through the action of PRC2, however, recent studies show that PRC1 is recruited to the same sites in the absence of PRC2.

Di and Trimethylation of H3 2ysine 9 (H3K9me2/3)

H3K9me2/3 is a well-characterised marker for heterochromatin, and is therefore strongly associated with gene repression. The formation of heterochromatin has been best studied in the yeast Schizosaccharomyces pombe, where it is initiated by recruitment of the RNA-induced transcriptional silencing (RITS) complex to double stranded RNAs produced from centromeric repeats. RITS recruits the Clr4 histone methyltransferase which deposits H3K9me2/3. This process is called histone methylation. H3K9Me2/3 serves as a binding site for the recruitment of Swi6 (heterochromatin protein 1 or HP1, another classic heterochromatin marker) which in turn recruits further repressive activities including histone modifiers such as histone deacetylases and histone methyltransferases.

Trimethylation of H4 Lysine 20 (H4K20me3)

This modification is tightly associated with heterochromatin, although its functional importance remains unclear. This mark is placed by the Suv4-20h methyltransferase, which is at least in part recruited by heterochromatin protein 1.

Bivalent Promoters

Analysis of histone modifications in embryonic stem cells (and other stem cells) revealed many gene promoters carrying both H3K4Me3 and H3K27Me3, in other words these promoters display both activating and repressing marks simultaneously.

This peculiar combination of modifications marks genes that are poised for transcription; they are not required in stem cells, but are rapidly required after differentiation into some lineages. Once the cell starts to differentiate, these bivalent promoters are resolved to either active or repressive states depending on the chosen lineage.

Other Functions

DNA Damage

Marking sites of DNA damage is an important function for histone modifications. It also protects DNA from getting destroyed by ultraviolet radiation of sun.

Phosphorylation of H2AX at Serine 139 (γH2AX)

Phosphorylated H2AX (also known as gamma H2AX) is a marker for DNA double strand breaks, and forms part of the response to DNA damage. H2AX is phosphorylated early after detection of DNA double strand break, and forms a domain extending many kilobases either side of the damage. Gamma H2AX acts as a binding site for the protein MDC1, which in turn recruits key DNA repair proteins (this complex topic is well reviewed in) and as such, gamma H2AX forms a vital part of the machinery that ensures genome stability.

Acetylation of H3 Lysine 56 (H3K56Ac)

H3K56Acx is required for genome stability. H3K56 is acetylated by the p300/Rtt109 complex, but is rapidly deacetylated around sites of DNA damage. H3K56 acetylation is also required to stabilise stalled replication forks, preventing dangerous replication fork collapses. Although in general mammals make far greater use of histone modifications than microorganisms, a major role of H3K56Ac in DNA replication exists only in fungi, and this has become a target for antibiotic development.

DNA Repair

Trimethylation of H3 Lysine 36 (H3K36me3)

H3K36me3 has the ability to recruit the MSH2-MSH6 (hMutSα) complex of the DNA mismatch repair pathway. Consistently, regions of the human genome with high levels of H3K36me3 accumulate less somatic mutations due to mismatch repair activity.

Chromosome Condensation

Phosphorylation of H3 at Serine 10

The mitotic kinase aurora B phosphorylates histone H3 at serine 10, triggering a cascade of changes that mediate mitotic chromosome condensation. Condensed chromosomes therefore stain very strongly for this mark, but H3S10 phosphorylation is also present at certain chromosome sites outside mitosis, for example in pericentric heterochromatin of cells during G2. H3S10 phosphorylation has also been linked to DNA damage caused by R-loop formation at highly transcribed sites.

Phosphorylation H2B at Serine 10/14

Phosphorylation of H2B at serine 10 (yeast) or serine 14 (mammals) is also linked to chromatin condensation, but for the very different purpose of mediating chromosome condensation during apoptosis. This mark is not simply a late acting bystander in apoptosis as yeast carrying mutations of this residue are resistant to hydrogen peroxide-induced apoptotic cell death.

Addiction

Epigenetic modifications of histone tails in specific regions of the brain are of central importance in addictions. Once particular epigenetic alterations occur, they appear to be long lasting "molecular scars" that may account for the persistence of addictions.

Cigarette smokers (about 15% of the US population) are usually addicted to nicotine. After 7 days of nicotine treatment of mice, acetylation of both histone H3 and histone H4 was increased at the FosB promoter in the nucleus accumbens of the brain, causing 61% increase in FosB expression. This would also increase expression of the splice variant Delta FosB. In the nucleus accumbens of the brain, Delta FosB functions as a "sustained molecular switch" and "master control protein" in the development of an addiction.

About 7% of the US population is addicted to alcohol. In rats exposed to alcohol for up to 5 days, there was an increase in histone 3 lysine 9 acetylation in the pronociceptin promoter in the brain amygdala complex. This acetylation is an activating mark for pronociceptin. The nociceptin/nociceptin opioid receptor system is involved in the reinforcing or conditioning effects of alcohol.

Methamphetamine addiction occurs in about 0.2% of the US population. Chronic methamphetamine use causes methylation of the lysine in position 4 of histone 3 located at the promoters of the *c-fos* and the *C-C chemokine receptor 2 (ccr2)* genes, activating those genes in the nucleus accumbens (NAc). c-fos is well known to be important in addiction. The *ccr2* gene is also important in addiction, since mutational inactivation of this gene impairs addiction.

Histone Synthesis

The first step of chromatin structure duplication is the synthesis of histone proteins: H1, H2A, H2B, H3, H4. These proteins are synthesized during S phase of the cell cycle. There are different mechanisms which contribute to the increase of histone synthesis.

Yeast

Yeast carry one or two copies of each histone gene, which are not clustered but rather scattered throughout chromosomes. Histone gene transcription is controlled by multiple gene regulatory proteins such as transcription factors which bind to histone promoter regions. In budding yeast, the candidate gene for activation of histone gene expression is SBF. SBF is a transcription factor that is activated in late G1 phase, when it dissociates from its repressor Whi5. This occurs when Whi5 is phosphorylated by Cdc8 which is a G1/S Cdk. Suppression of histone gene expression outside of S phases is dependent on Hir proteins which form inactive chromatin structure at the locus of histone genes, causing transcriptional activators to be blocked.

Metazoans

In metazoans the increase in the rate of histone synthesis is due to the increase in processing of pre-mRNA to its mature form as well as decrease in mRNA degradation; this results in an increase of active mRNA for translation of histone proteins. The mechanism for mRNA activation has been found to be the removal of a segment of the 3' end of the mRNA strand, and is dependent on association with stem-loop binding protein (SLBP). SLBP also stabilizes histone mRNAs during S phase by blocking degradation by the 3'hExo nuclease. SLBP levels are controlled by cell-cycle proteins, causing SLBP to accumulate as cells enter S phase and degrade as cells leave S phase. SLBP are marked for degradation by phosphorylation at two threonine residues by cyclin dependent kinases, possibly cyclin A/ cdk2, at the end of S phase. Metazoans also have multiple copies of histone genes clustered on chromosomes which are localized in structures called Cajal bodies as determined by genome-wide chromosome conformation capture analysis (4C-Seq).

Link between Cell-cycle Control Machinery and Histone Synthesis

Nuclear protein Ataxia-Telangiectasia (NPAT), also known as nuclear protein coactivator of histone transcription, is a transcription factor which activates histone gene transcription on chromosomes 1 and 6 of human cells. NPAT is also a substrate of cyclin E-Cdk2, which is required for the transition between G1 phase and S phase. NPAT activates histone gene expression only after it has been phosphorylated by the G1/S-Cdk cyclin E-Cdk2 in early S phase. This shows an important regulatory link between cell-cycle control and histone synthesis.

Funtions of DNA-binding Protein

DNA-binding proteins have diverse functions in the cell, and play vital roles in various cellular processes, such as gene regulation, DNA replication, and repair. Identification of DNA-binding proteins is one of the most important tasks in the annotation of protein functions. In recent years, DNA-binding proteins can be identified by several experimental techniques, including filter binding assays, X-ray crystallography3 and NMR4. However, it is time-consuming and expensive to identify DNA-binding proteins by experimental approaches. Facing the avalanche of new protein sequences generated in the post-genomic and big data age it is highly desired to develop automated methods for rapidly and effectively identifying DNA-binding proteins basing on the protein sequence information alone.

The computational methods for DNA binding protein identification can be grouped into two categories: (i) methods based on structures (ii) methods based on sequences. The first type makes use of both the structural and sequential information of target proteins. Although these methods show promising predictive performance, the structural information of proteins is not always available, particularly for the huge amount of proteins, which prevents the application of these methods. In contrast, the second type methods overcome this shortcoming by only requiring the sequence information as input for the prediction.

A key to improve the performance of the sequence-based methods is to find suitable feature extraction algorithms that can capture the characteristics of DNA binding proteins and non DNA binding proteins. Motivated by the successful application of Chou's pseudo amino acid composition (PseAAC) to many important tasks in the field of computational proteomics, here we are to propose a new approach for DNA binding protein identification called iDNAPro-PseAAC, which extends the classic PseAAC approach by incorporating the evolutionary information in the form of profile-based protein representation. The iDNAPro-PseAAC has the following advantages compared with other currently available approaches: (i) It is able to incorporate the global or long range sequence-order effects by means of PseAAC. (ii) The evolutionary information imbedded in the profile-based protein representation is employed by iDNAPro-PseAAC. (iii) It considers the various physical-chemical properties of amino acids.

To establish a really useful statistical predictor for a protein system, we need to consider the following procedures: (i) Construct or select a valid benchmark dataset to train and test the predictor. (ii) Formulate the protein samples with an effective mathematical expression that can truly reflect their intrinsic correlation with the attribute to be predicted. (iii) Introduce or develop a powerful algorithm (or engine) to operate the prediction. (iv) Perform properly cross-validation tests to objectively evaluate the anticipated accuracy of the predictor.

References

- Imperfect DNA replication results in mutations. Berg JM, Tymoczko JL, Stryer L, Clarke ND (2002). "Chapter 27: DNA Replication, Recombination, and Repair". Biochemistry. W.H. Freeman and Company. ISBN 0-7167-3051-0

- Frick, David; Richardson, Charles (July 2001). "DNA Primases". Annual Review of Biochemistry. 70: 39–80. Doi:10.1146/annurev.biochem.70.1.39. PMID 11395402
- Watson J, Baker T, Bell S, Gann A, Levine M, Losick R, Molecular Biology of the Gene 6th Edition, Pearson Education Inc, 2008. ISBN 080539592X
- Dame RT (2005). "The role of nucleoid-associated proteins in the organization and compaction of bacterial chromatin". Mol. Microbiol. 56 (4): 858–70. Doi:10.1111/j.1365-2958.2005.04598.x. PMID 15853876
- Cox M, Nelson DR, Lehninger AL (2005). Lehninger Principles of Biochemistry. San Francisco: W.H. Freeman. ISBN 978-0-7167-4339-2
- Ozdemir A, Spicuglia S, Lasonder E, Vermeulen M, Campsteijn C, Stunnenberg HG, Logie C (Jul 2005). "Characterization of lysine 56 of histone H3 as an acetylation site in Saccharomyces cerevisiae". The Journal of Biological Chemistry. 280 (28): 25949–52. Doi:10.1074/jbc.C500181200. PMID 15888442

4

DNA Damage and Repair

The alteration in the chemical structure of DNA which causes changes in the structure of genetic material is referred to as DNA damage. DNA repair refers to the processes by which a cell identifies and cures damage to the DNA molecules which encode its genome. The topics elaborated in this chapter will help in gaining a better perspective about DNA damage and repair.

DNA Damage

The most significant consequence of oxidative stress in the body is thought to be damage to DNA. DNA may be modified in a variety of ways, which can ultimately lead to mutations and genomic instability. This could result in the development of a variety of cancers including colon, breast, and prostate.

Oxidative DNA damage refers to the oxidation of specific bases. 8-hydroxydeoxyguanosine (8-OHdG) is the most common marker for oxidative DNA damage and can be measured in virtually any species. It is formed and enhanced most often by chemical carcinogens. A similar oxidative damage can occur in RNA with the formation of 8-OHG (8-hydroxyguanosine), which has been implicated in various neurological disorders.

Hydrolytic DNA damage involves deamination or the total removal of individual bases. Loss of DNA bases, known as AP (apurinic/apyrimidinic) sites, can be particularly mutagenic and if left unrepaired they can inhibit transcription. Hydrolytic damage may result from the biochemical reactions of various metabolites as well as the overabundance of reactive oxygen species.

Ultraviolet and other types of radiation can damage DNA in the form of DNA strand breaks. This involves a cut in one or both DNA strands; double-strand breaks are especially dangerous and can be mutagenic, since they can potentially affect the expression of multiple genes. UV-induced damage can also result in the production of pyrimidine dimers, where covalent cross-links occur in cytosine and thymine residues. The most common pyrimidine dimers are cyclobutane pyrimidine dimers (CPD) and pyrimidine (6-4) pyrimidone photoproducts (6-4PP). CPD and 6-4PP are the most frequent DNA mutations found in the p53 protein in skin cancers. Pyrimidine dimers can disrupt polymerases and prevent proper replication of DNA.

DNA damage may also result from exposure to polycyclic aromatic hydrocarbons (PAHs). PAHs are potent, ubiquitous atmospheric pollutants commonly associated with oil, coal, cigarette smoke, and automobile exhaust fumes. A common marker for DNA damage due to PAHs is Benzo(a)pyrene diol epoxide (BPDE). BPDE is found to be very reactive, and known to bind covalently to proteins, lipids, and guanine residues of DNA to produce BPDE adducts. If left unrepaired, BPDE-DNA adducts may lead to permanent mutations resulting in cell transformation and ultimately tumor development.

The Comet Assay, or single cell gel electrophoresis assay (SCGE), is a common technique used to measure all types of DNA damage, including the various types of damage. It is a convenient tool for measuring universal DNA damage in individual cells.

DNA Damage Response

The packaging of eukaryotic DNA into chromatin is a barrier to all DNA-based processes that require enzyme action. For most DNA repair processes, the chromatin must be remodeled. In eukaryotes, ATP-dependent chromatin remodeling complexes and histone-modifying enzymes are two factors that act to accomplish this remodeling process after DNA damage occurs. Further DNA repair steps, involving multiple enzymes, usually follow.

Base Excision Repair

Oxidized bases in DNA are produced in cells treated with Hoechst dye followed by micro-irradiation with 405 nm light. Such oxidized bases can be repaired by base excision repair.

When the 405 nm light is focused along a narrow line within the nucleus of a cell, about 2.5 seconds after irradiation, the chromatin remodeling enzyme Alc1 achieves half-maximum recruitment onto the irradiated micro-line. The line of chromatin that was irradiated then relaxes, expanding side-to-side over the next 60 seconds.

Within 6 seconds of the irradiation with 405 nm light, there is half-maximum recruitment of OGG1 to the irradiated line. OGG1 is an enzyme that removes the oxidative DNA damage 8-oxo-dG from DNA. Removal of 8-oxo-dG, during base excision repair, occurs with a half-life of 11 minutes.

Nucleotide Excision Repair

Ultraviolet (UV) light induces the formation of DNA damages including pyrimidine dimers (such as thymine dimers) and 6,4 photoproducts. These types of "bulky" damages are repaired by nucleotide excision repair.

After irradiation with UV light, DDB2, in a complex with DDB1, the ubiquitin ligase protein CUL4A and the RING finger protein ROC1, associates with sites of damage within chromatin. Half-maximum association occurs in 40 seconds. PARP1 also associates within this period. The PARP1 protein attaches to both DDB1 and DDB2 and then PARylates (creates a poly-ADP ribose chain) on DDB2 that attracts the DNA remodeling protein ALC1. ALC1 relaxes chromatin at sites of UV damage to DNA. In addition, the ubiquitin E3 ligase complex DDB1-CUL4A carries out ubiquitination of the core histones H2A, H3, and H4, as well as the repair protein XPC, which has been attracted to the site of the DNA damage. XPC, upon ubiquitination, is activated and initiates the nucleotide excision repair pathway. Somewhat later, at 30 minutes after UV damage, the INO80 chromatin remodeling complex is recruited to the site of the DNA damage, and this coincides with the binding of further nucleotide excision repair proteins, including ERCC1.

Homologous Recombinational Repair

Double-strand breaks (DSBs) at specific sites can be induced by transfecting cells with a plasmid encoding I-SceI endonuclease (a homing endonuclease). Multiple DSBs can be induced by irradiating sensitized cells (labeled with 5'-bromo-2'-deoxyuridine and with Hoechst dye) with 780 nm light. These DSBs can be repaired by the accurate homologous recombinational repair or by the less accurate non-homologous end joining repair pathway. Here we describe the early steps in homologus recombinational repair (HRR).

After treating cells to introduce DSBs, the stress-activated protein kinase, c-Jun N-terminal kinase (JNK), phosphorylates SIRT6 on serine 10. This post-translational modification facilitates the mobilization of SIRT6 to DNA damage sites with half-maximum recruitment in well under a second. SIRT6 at the site is required for efficient recruitment of poly (ADP-ribose) polymerase 1 (PARP1) to a DNA break site and for efficient repair of DSBs. PARP1 protein starts to appear at DSBs in less than a second, with half maximum accumulation within 1.6 seconds after the damage occurs. This then allows half maximum recruitment of the DNA repair enzymes MRE11 within 13 seconds and NBS1 within 28 seconds. MRE11 and NBS1 carry out early steps of the HRR pathway.

γH2AX, the phosphorylated form of H2AX is also involved in early steps of DSB repair. The histone variant H2AX constitutes about 10% of the H2A histones in human chromatin. γH2AX (H2AX phosphorylated on serine 139) can be detected as soon as 20 seconds after irradiation of cells (with DNA double-strand break formation), and half maximum accumulation of γH2AX occurs in one minute. The extent of chromatin with phosphorylated γH2AX is about two million base pairs at the site of a DNA double-strand break. γH2AX does not, itself, cause chromatin decondensation, but within 30 seconds of irradiation, RNF8 protein can be detected in association with γH2AX. RNF8 mediates extensive chromatin decondensation, through its subsequent interaction with CHD4, a component of the nucleosome remodeling and deacetylase complex NuRD.

DNA Damage in Memory Formation

Oxidation of Guanine

Oxidation of guanine, particularly within CpG sites, may be especially important in learning and memory. Methylation of cytosines occurs at 60–90% of CpG sites depending on the tissue type. In the mammalian brain, ~62% of CpGs are methylated. Methylation of CpG sites tends to stably silence genes. More than 500 of these CpG sites are de-methylated in neuron DNA during memory formation and memory consolidation in the hippocampus and cingulate cortex regions of the brain. As indicated below, the first step in de-methylation of methylated cytosine at a CpG site is oxidation of the guanine to form 8-oxo-dG.

Role of Oxidized Guanine in DNA De-methylation

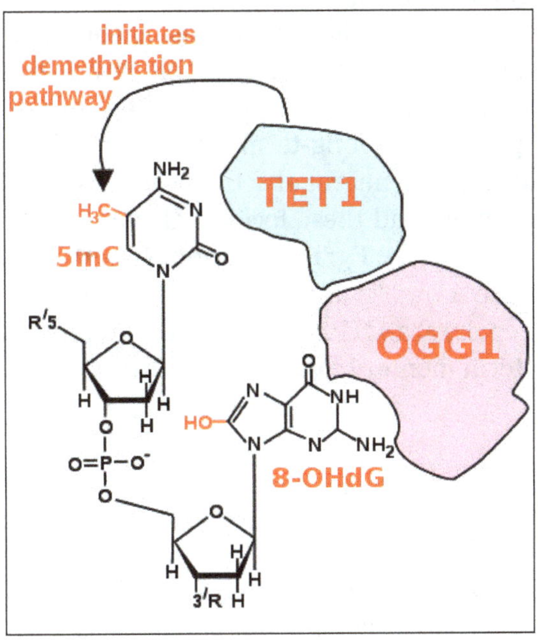

Initiation of DNA demethylation at a CpG site. In adult somatic cells DNA methylation typically occurs in the context of CpG dinucleotides (CpG sites), forming 5-methylcytosine-pG, or 5mCpG. Reactive oxygen species (ROS) may attack guanine at the dinucleotide site, forming 8-hydroxy-2'-deoxyguanosine (8-OHdG), and resulting in a 5mCp-8-OHdG dinucleotide site. The base excision repair enzyme OGG1 targets 8-OHdG and binds to the lesion without immediate excision. OGG1, present at a 5mCp-8-OHdG site recruits TET1 and TET1 oxidizes the 5mC adjacent to the 8-OHdG. This initiates demethylation of 5mC.

The figure here shows a CpG site where the cytosine is methylated to form 5-methylcytosine (5mC) and the guanine is oxidized to form 8-oxo-2'-deoxyguanosine (in the figure this is shown in the tautomeric form 8-OHdG). When this structure is formed, the base excision repair enzyme OGG1 targets 8-OHdG and binds to the lesion without immediate excision. OGG1, present at a 5mCp-8-OHdG site recruits TET1, and TET1 oxidizes the 5mC adjacent to the 8-OHdG. This initiates de-methylation of 5mC. TET1 is a key enzyme involved in de-methylating 5mCpG. However, TET1 is only able to act on 5mCpG if the guanine was first oxidized to form 8-hydroxy-2'-deoxyguanosine (8-OHdG or its tautomer 8-oxo-dG), resulting in a 5mCp-8-OHdG dinucleotide. This initiates the de-methylation pathway on the methylated cytosine, finally resulting in an unmethylated cytosine.

Altered protein expression in neurons, due to changes in methylation of DNA, (likely controlled by 8-oxo-dG-dependent de-methylation of CpG sites in gene promoters within neuron DNA) has been established as central to memory formation.

Role of Double-strand Breaks in Memory Formation

Exposure of mice to physiological learning behaviors in vivo, such as being exposed to a new environment or activation of the primary visual cortex (V1) by exposing mice to visual stimuli, results in the formation of DNA double-strand breaks (DSBs) in the dentate gyrus (part of the hippocampus brain region). Similarly, exposing mice to contextual fear conditioning, producing a long-term memory, also causes DSBs in the hippocampus within 15 minutes. These neuronal activity-induced DSBs are restricted to only 21 loci in the neuron genome, and these loci are also enriched for the early response genes (including Fos, FosB, Npas4, Egr1, Nr4a1 and Nr4a3) in neuron activation. These neural activity-induced DNA breaks are generated by a Type II topoisomerase. An inhibitor of NHEJ DSB repair, ara-CTP, (likely inhibiting repair of such double-strand breaks), prevents long-term memory formation.

DNA Repair

DNA repair is a collection of processes by which a cell identifies and corrects damage to the DNA molecules that encode its genome. In human cells, both normal metabolic activities and environmental factors such as radiation can cause DNA damage, resulting in as many as 1 million individual molecular lesions per cell per day. Many of these lesions cause structural damage to the DNA molecule and can alter or eliminate the cell's ability to transcribe the gene that the affected DNA encodes. Other lesions induce potentially harmful mutations in the cell's genome, which affect the survival of its daughter cells after it undergoes mitosis. As a consequence, the DNA repair process is constantly active as it responds to damage in the DNA structure. When normal repair processes fail, and when cellular apoptosis does not occur, irreparable DNA damage may occur, including double-strand breaks and DNA crosslinkages (interstrand crosslinks or ICLs). This can eventually lead to malignant tumors, or cancer as per the two hit hypothesis.

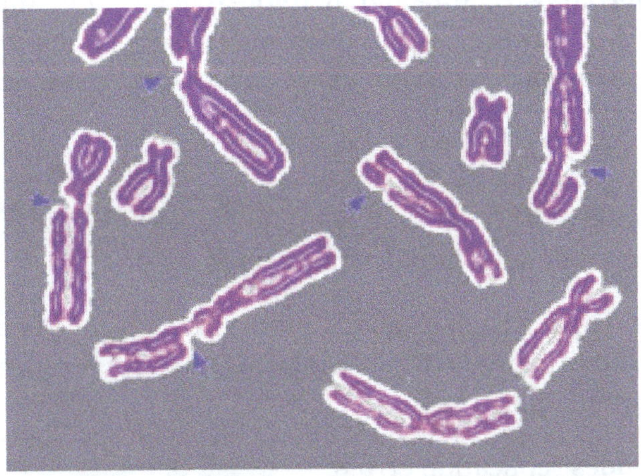

DNA damage resulting in multiple broken chromosomes.

The rate of DNA repair is dependent on many factors, including the cell type, the age of the cell, and the extracellular environment. A cell that has accumulated a large amount of DNA damage, or one that no longer effectively repairs damage incurred to its DNA, can enter one of three possible states:

- An irreversible state of dormancy, known as senescence.
- Cell suicide, also known as apoptosis or programmed cell death.
- Unregulated cell division, which can lead to the formation of a tumor that is cancerous.

The DNA repair ability of a cell is vital to the integrity of its genome and thus to the normal functionality of that organism. Many genes that were initially shown to influence life span have turned out to be involved in DNA damage repair and protection.

The 2015 Nobel Prize in Chemistry was awarded to Tomas Lindahl, Paul Modrich, and Aziz Sancar for their work on the molecular mechanisms of DNA repair processes.

DNA Damage

DNA damage, due to environmental factors and normal metabolic processes inside the cell, occurs at a rate of 10,000 to 1,000,000 molecular lesions per cell per day. While this constitutes only 0.000165% of the human genome's approximately 6 billion bases (3 billion base pairs), unrepaired lesions in critical genes (such as tumor suppressor genes) can impede a cell's ability to carry out its function and appreciably increase the likelihood of tumor formation and contribute to tumour heterogeneity.

The vast majority of DNA damage affects the primary structure of the double helix; that is, the bases themselves are chemically modified. These modifications can in turn disrupt the molecules' regular helical structure by introducing non-native chemical bonds or bulky adducts that do not fit in the standard double helix. Unlike proteins and RNA, DNA usually lacks tertiary structure and therefore damage or disturbance does not occur at that level. DNA is, however, supercoiled and wound around "packaging" proteins called histones (in eukaryotes), and both superstructures are vulnerable to the effects of DNA damage.

Sources of DNA

DNA damage can be subdivided into two main types:

- Endogenous damage such as attack by reactive oxygen species produced from normal metabolic byproducts (spontaneous mutation), especially the process of oxidative deamination.
    - Also includes replication errors.
- Exogenous damage caused by external agents such as:
    - Ultraviolet [UV 200–400 nm] radiation from the sun.
    - Other radiation frequencies, including x-rays and gamma rays.
    - Hydrolysis or thermal disruption.
    - Certain plant toxins.
    - Human-made mutagenic chemicals, especially aromatic compounds that act as DNA intercalating agents.
    - Viruses.

The replication of damaged DNA before cell division can lead to the incorporation of wrong bases opposite damaged ones. Daughter cells that inherit these wrong bases carry mutations from which the original DNA sequence is unrecoverable (except in the rare case of a back mutation, for example, through gene conversion).

Types

There are several types of damage to DNA due to endogenous cellular processes:

- Oxidation Of bases [e.g. 8-oxo-7,8-dihydroguanine (8-oxoG)] and generation of DNA strand interruptions from reactive oxygen species.
- Alkylation of bases (usually methylation), such as formation of 7-methylguanosine, 1-methyladenine, 6-O-Methylguanine.

- Hydrolysis of bases, such as deamination, depurination, and depyrimidination.
- "Bulky adduct formation" (e.g., benzo[a]pyrene diol epoxide-dG adduct, aristolactam I-dA adduct).
- Mismatch of bases, due to errors in DNA replication, in which the wrong DNA base is stitched into place in a newly forming DNA strand, or a DNA base is skipped over or mistakenly inserted.
- Monoadduct damage cause by change in single nitrogenous base of DNA.
- Diadduct damage.

Damage caused by exogenous agents comes in many forms. Some examples are:

- UV-B light causes crosslinking between adjacent cytosine and thymine bases creating pyrimidine dimers. This is called direct DNA damage.
- UV-A light creates mostly free radicals. The damage caused by free radicals is called indirect DNA damage.
- Ionizing radiation such as that created by radioactive decay or in cosmic rays causes breaks in DNA strands. Intermediate-level ionizing radiation may induce irreparable DNA damage (leading to replicational and transcriptional errors needed for neoplasia or may trigger viral interactions) leading to pre-mature aging and cancer.
- Thermal disruption at elevated temperature increases the rate of depurination (loss of purine bases from the DNA backbone) and single-strand breaks. For example, hydrolytic depurination is seen in the thermophilic bacteria, which grow in hot springs at 40–80 °C. The rate of depurination (300 purine residues per genome per generation) is too high in these species to be repaired by normal repair machinery, hence a possibility of an adaptive response cannot be ruled out.
- Industrial chemicals such as vinyl chloride and hydrogen peroxide, and environmental chemicals such as polycyclic aromatic hydrocarbons found in smoke, soot and tar create a huge diversity of DNA adducts- ethenobases, oxidized bases, alkylated phosphotriesters and crosslinking of DNA, just to name a few.

UV damage, alkylation/methylation, X-ray damage and oxidative damage are examples of induced damage. Spontaneous damage can include the loss of a base, deamination, sugar ring puckering and tautomeric shift. Constitutive (spontaneous) DNA damage caused by endogenous oxidants can be detected as a low level of histone H2AX phosphorylation in untreated cells.

Nuclear versus Mitochondrial

In human cells, and eukaryotic cells in general, DNA is found in two cellular locations – inside the nucleus and inside the mitochondria. Nuclear DNA (nDNA) exists as chromatin during non-replicative stages of the cell cycle and is condensed into aggregate structures known as chromosomes during cell division. In either state the DNA is highly compacted and wound up around bead-like proteins called histones. Whenever a cell needs to express the genetic information encoded in its nDNA the required chromosomal region is unravelled, genes located therein are expressed, and then the region is condensed back to its resting conformation. Mitochondrial DNA (mtDNA) is located inside mitochondria organelles, exists in multiple copies, and is also tightly associated with a number of proteins to form a complex known as the nucleoid. Inside mitochondria, reactive oxygen species (ROS), or free radicals, byproducts of the constant production of adenosine triphosphate (ATP) via oxidative phosphorylation, create a highly oxidative environment that is known to damage mtDNA. A critical enzyme in counteracting the toxicity of these species is superoxide dismutase, which is present in both the mitochondria and cytoplasm of eukaryotic cells.

Senescence and Apoptosis

Senescence, an irreversible process in which the cell no longer divides, is a protective response to the shortening of the chromosome ends. The telomeres are long regions of repetitive noncoding DNA that cap chromosomes and undergo partial degradation each time a cell undergoes division. In contrast, quiescence is a reversible state of cellular dormancy that is unrelated to genome damage. Senescence in cells may serve as a functional alternative to apoptosis in cases where the physical presence of a cell for spatial reasons is required by the organism, which serves as a "last resort" mechanism to prevent a cell with damaged DNA from replicating inappropriately in the absence of pro-growth cellular signaling. Unregulated cell division can lead to the formation of a tumor, which is potentially lethal to an organism. Therefore, the induction of senescence and apoptosis is considered to be part of a strategy of protection against cancer.

Mutation

It is important to distinguish between DNA damage and mutation, the two major types of error in DNA. DNA damage and mutation are fundamentally different. Damage results in physical abnormalities in the DNA, such as single- and double-strand breaks, 8-hydroxydeoxyguanosine residues, and polycyclic aromatic hydrocarbon adducts. DNA damage can be recognized by enzymes, and thus can be correctly repaired if redundant information, such as the undamaged sequence in the complementary DNA strand or in a homologous chromosome, is available for copying. If a cell retains DNA damage, transcription of a gene can be prevented, and thus translation into a protein will also be blocked. Replication may also be blocked or the cell may die.

In contrast to DNA damage, a mutation is a change in the base sequence of the DNA. A mutation cannot be recognized by enzymes once the base change is present in both DNA strands, and thus a mutation cannot be repaired. At the cellular level, mutations can cause alterations in protein function and regulation. Mutations are replicated when the cell replicates. In a population of cells, mutant cells will increase or decrease in frequency according to the effects of the mutation on the ability of the cell to survive and reproduce.

Although distinctly different from each other, DNA damage and mutation are related because DNA damage often causes errors of DNA synthesis during replication or repair; these errors are a major source of mutation.

Given these properties of DNA damage and mutation, it can be seen that DNA damage is a special problem in non-dividing or slowly-dividing cells, where unrepaired damage will tend to accumulate over time. On the other hand, in rapidly-dividing cells, unrepaired DNA damage that does not kill the cell by blocking replication will tend to cause replication errors and thus mutation. The great majority of mutations that are not neutral in their effect are deleterious to a cell's survival. Thus, in a population of cells composing a tissue with replicating cells, mutant cells will tend to be lost. However, infrequent mutations that provide a survival advantage will tend to clonally expand at the expense of neighboring cells in the tissue. This advantage to the cell is disadvantageous to the whole organism, because such mutant cells can give rise to cancer. Thus, DNA damage in frequently dividing cells, because it gives rise to mutations, is a prominent cause of cancer. In contrast, DNA damage in infrequently-dividing cells is likely a prominent cause of aging.

Mechanisms

Cells cannot function if DNA damage corrupts the integrity and accessibility of essential information in the genome (but cells remain superficially functional when non-essential genes are missing or damaged). Depending on the type of damage inflicted on the DNA's double helical structure, a variety of repair strategies have evolved to restore lost information. If possible, cells use the unmodified complementary strand of the DNA or the sister chromatid as a template to recover the original information. Without access to a template, cells use an error-prone recovery mechanism known as translesion synthesis as a last resort.

Damage to DNA alters the spatial configuration of the helix, and such alterations can be detected by the cell. Once damage is localized, specific DNA repair molecules bind at or near the site of damage, inducing other molecules to bind and form a complex that enables the actual repair to take place.

Direct Reversal

Cells are known to eliminate three types of damage to their DNA by chemically reversing it. These mechanisms do not require a template, since the types of damage they counteract can occur in only one of the four bases. Such direct reversal mechanisms are specific to the type of damage incurred and do not involve breakage of the phosphodiester backbone. The formation of pyrimidine dimers upon irradiation with UV light results in an abnormal covalent bond between adjacent pyrimidine bases. The photoreactivation process directly reverses this damage by the action of the enzyme photolyase, whose activation is obligately dependent on energy absorbed from blue/UV light (300–500 nm wavelength) to promote catalysis. Photolyase, an old enzyme present in bacteria, fungi, and most animals no longer functions in humans, who instead use nucleotide excision repair to repair damage from UV irradiation. Another type of damage, methylation of guanine bases, is directly reversed by the protein methyl guanine methyl transferase (MGMT), the bacterial equivalent of which is called ogt. This is an expensive process because each MGMT molecule can be used only once; that is, the reaction is stoichiometric rather than catalytic. A generalized response to methylating agents in bacteria is known as the adaptive response and confers a level of resistance to alkylating agents upon sustained exposure by upregulation of alkylation repair enzymes. The third type of DNA damage reversed by cells is certain methylation of the bases cytosine and adenine.

Mismatch Repaired by DNA Polymerase Proofreading

The frequency of base pair substitution mutations in bacterial genes ranges from 10^{-7} to 10^{-11} errors/ generations. However DNA polymerase inserts incorrect nucleotide at frequency of 10^{-5}. Most of the difference between the two values is accounted for by the 3' to 5' exonuclease proofreading activity of the DNA polymerase in both bacteria and eukaryotes. When an incorrect nucleotide is inserted, the polymerase detects the mismatched base pairs and corrects the area by back spacing to remove the wrong nucleotide and then resuming synthesis in the forward directions.

The mutator mutations in *E.coli* illustrate importance of the 3' to 5' exonuclease activity of the DNA polymerase for maintaining low mutation rate. Mutator mutants have much higher than normal mutation frequency for all genes. These mutants have mutations in genes for proteins whose normal functions are required for accurate DNA replication. For example mut D mutator gene of *E.coli* encodes the epsilon sub-unit of DNA polymerase III, the primary replication enzyme of *E.coli*. The mut D mutants are defective in 3' to 5' proofreading activity, so that many incorrectly inserted nucleotide are left unrepaired.

Repair of UV Induced Pyrimidine Dimers

Through photo reactivation or light repair, UV light induced thymine (other pyrimidines) dimmers are reverted directly to the original form by exposure to near-UV light in the wavelength range from 320 to 370 nm. Photo reactivation occurs when an enzyme called photolyase is activated by a photon of light and splits the dimmers apart. strains with mutations in the phr gene are defective in light repair. Photolyase has been found in bacteria and in simple eukaryotes but not in humans.

Repair of Alkylation Damage

Alkylating agents transfer alkyl groups (usually methyl or ethyl groups) onto the bases. The mutagen MMS methylates the oxygen of carbon 6 in guanine. In *E. coli* this alkylation damage is repaired by an enzyme called O^6 methylguanine methyltransferase. This enzyme removes methyl groups from guanine, thereby changing the base back to its original form. Similar specific systems exist to repair alkylated thymines. Mutations of the genes encoding this repair enzymes results in a much higher rate of spontaneous mutation.

Methyl Directed Mismatch Repair

Despite proofreading of DNA polymerase, a number of mismatched base pairs remain uncorrected after replication has been completed. In the next round of replication these errors will become fixed as mutations if they are not repaired. This system recognizes mismatched base pairs, excises the incorrect base, and then carries out repair synthesis.

Single-strand Damage

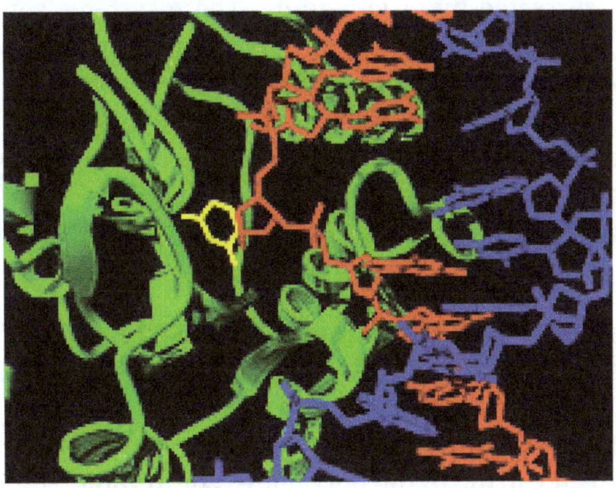

Structure of the base-excision repair enzyme uracil-DNA glycosylase excising a hydrolytically-produced uracil residue from DNA. The uracil residue is shown in yellow.

When only one of the two strands of a double helix has a defect, the other strand can be used as a template to guide the correction of the damaged strand. In order to repair damage to one of the two paired molecules of DNA, there exist a number of excision repair mechanisms that remove the damaged nucleotide and replace it with an undamaged nucleotide complementary to that found in the undamaged DNA strand.

- Base excision repair (BER) damaged single bases or nucleotide are most commonly repaired by removing the base or the nucleotide involved and then inserting the correct base or nucleotide. In base excision repair, repair glycosylases enzyme removes the damaged base from the DNA by cleaving the bond between base and deoxyribose sugars. These enzymes remove a single nitrogenous base to create an apurinic or apyrimidinic site (AP site). Enzymes called AP endonucleases nick the damaged DNA backbone at the AP site. DNA polymerase then removes the damaged region using its 5' to 3' exonuclease activity and correctly synthesizes the new strand using the complementary strand as a template. The gap is then sealed by enzyme DNA ligase.

- Nucleotide excision repair (NER) repairs damaged DNA which commonly consists of bulky, helix-distorting damage, such as pyrimidine dimerization caused by UV light. Damaged regions are removed in 12–24 nucleotide-long strands in a three-step process which consists of recognition of damage, excision of damaged DNA both upstream and downstream of damage by endonucleases, and resynthesis of removed DNA region. NER is a highly evolutionarily conserved repair mechanism and is used in nearly all eukaryotic and prokaryotic cells. In prokaryotes, NER is mediated by Uvr proteins. In eukaryotes, many more proteins are involved, although the general strategy is the same.
- Mismatch repair systems are present in essentially all cells to correct errors that are not corrected by proofreading. These systems consist of at least two proteins. One detects the mismatch, and the other recruits an endonuclease that cleaves the newly synthesized DNA strand close to the region of damage. In *E. coli* , the proteins involved are the Mut class proteins. This is followed by removal of damaged region by an exonuclease, resynthesis by DNA polymerase, and nick sealing by DNA ligase.

Double-strand Breaks

Double-strand breaks, in which both strands in the double helix are severed, are particularly hazardous to the cell because they can lead to genome rearrangements. It was noted in some studies that double-strand breaks and a "cross-linkage joining both strands at the same point is irreparable because neither strand can then serve as a template for repair. The cell will die in the next mitosis or in some rare instances, mutate." Three mechanisms exist to repair double-strand breaks (DSBs): non-homologous end joining (NHEJ), microhomology-mediated end joining (MMEJ), and homologous recombination (HR). In an *in vitro* system, MMEJ occurred in mammalian cells at the levels of 10–20% of HR when both HR and NHEJ mechanisms were also available.

In NHEJ, DNA Ligase IV, a specialized DNA ligase that forms a complex with the cofactor XRCC4, directly joins the two ends. To guide accurate repair, NHEJ relies on short homologous sequences called microhomologies present on the single-stranded tails of the DNA ends to be joined. If these overhangs are compatible, repair is usually accurate. NHEJ can also introduce mutations during repair. Loss of damaged nucleotides at the break site can lead to deletions, and joining of nonmatching termini forms insertions or translocations. NHEJ is especially important before the cell has replicated its DNA, since there is no template available for repair by homologous recombination. There are "backup" NHEJ pathways in higher eukaryotes. Besides its role as a genome caretaker, NHEJ is required for joining hairpin-capped double-strand breaks induced during V(D)J recombination, the process that generates diversity in B-cell and T-cell receptors in the vertebrate immune system.

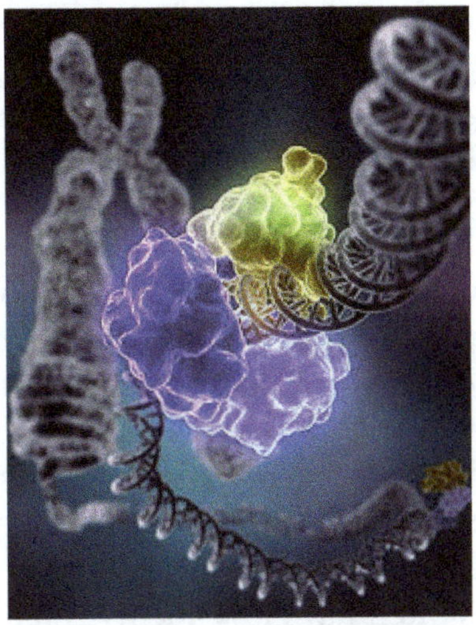

DNA ligase, repairing chromosomal damage, is an enzyme that joins broken nucleotides together by catalyzing the formation of an internucleotide ester bond between the phosphate backbone and the deoxyribose nucleotides.

Homologous recombination requires the presence of an identical or nearly identical sequence to be used as a template for repair of the break. The enzymatic machinery responsible for this repair process is nearly identical to the machinery responsible for chromosomal crossover during meiosis. This pathway allows a damaged chromosome to be repaired using a sister chromatid (available in G2 after DNA replication) or a homologous chromosome as a template. DSBs caused by the replication machinery attempting to synthesize across a single-strand break or unrepaired lesion cause collapse of the replication fork and are typically repaired by recombination.

MMEJ starts with short-range end resection by MRE11 nuclease on either side of a double-strand break to reveal microhomology regions. In further steps, Poly (ADP-ribose) polymerase 1 (PARP1) is required and may be an early step in MMEJ. There is pairing of microhomology regions followed by recruitment of flap structure-specific endonuclease 1 (FEN1) to remove overhanging flaps. This is followed by recruitment of XRCC1–LIG3 to the site for ligating the DNA ends, leading to an intact DNA. MMEJ is always accompanied by a deletion, so that MMEJ is a mutagenic pathway for DNA repair.

The extremophile *Deinococcus radiodurans* has a remarkable ability to survive DNA damage from ionizing radiation and other sources. At least two copies of the genome, with random DNA breaks, can form DNA fragments through annealing. Partially overlapping fragments are then used for synthesis of homologous regions through a moving D-loop that can continue extension until they find complementary partner strands. In the final step there is crossover by means of RecA-dependent homologous recombination.

Topoisomerases introduce both single- and double-strand breaks in the course of changing the DNA's state of supercoiling, which is especially common in regions near an open replication fork. Such breaks are not considered DNA damage because they are a natural intermediate in the topoisomerase biochemical mechanism and are immediately repaired by the enzymes that created them.

Translesion Synthesis

Translesion synthesis (TLS) is a DNA damage tolerance process that allows the DNA replication machinery to replicate past DNA lesions such as thymine dimers or AP sites. It involves switching out regular DNA polymerases for specialized translesion polymerases (i.e. DNA polymerase IV or V, from the Y Polymerase family), often with larger active sites that can facilitate the insertion of bases opposite damaged nucleotides. The polymerase switching is thought to be mediated by, among other factors, the post-translational modification of the replication processivity factor PCNA. Translesion synthesis polymerases often have low fidelity (high propensity to insert wrong bases) on undamaged templates relative to regular polymerases. However, many are extremely efficient at inserting correct bases opposite specific types of damage. For example, Pol η mediates error-free bypass of lesions induced by UV irradiation, whereas Pol ι introduces mutations at these sites. Pol η is known to add the first adenine across the T^T photodimer using Watson-Crick base pairing and the second adenine will be added in its syn conformation using Hoogsteen base pairing. From a cellular perspective, risking the introduction of point mutations during translesion synthesis may be preferable to resorting to more drastic mechanisms of DNA repair, which may cause gross chromosomal aberrations or cell death. In short, the process involves specialized polymerases either bypassing or repairing lesions at locations of stalled DNA replication. For example, Human DNA polymerase eta can bypass complex DNA lesions like guanine-thymine intra-strand crosslink, G[8,5-Me]T, although it can cause targeted and semi-targeted mutations. Paromita Raychaudhury and Ashis Basu studied the toxicity and mutagenesis of the same lesion in *Escherichia coli* by replicating a G[8,5-Me] T-modified plasmid in *E. coli* with specific DNA polymerase knockouts. Viability was very low in a strain lacking pol II, pol IV, and pol V, the three SOS-inducible DNA polymerases, indicating that translesion synthesis is conducted primarily by these specialized DNA polymerases. A bypass platform is provided to these polymerases by Proliferating cell nuclear antigen (PCNA). Under normal circumstances, PCNA bound to polymerases replicates the DNA. At a site of lesion, PCNA is ubiquitinated, or modified, by the RAD6/RAD18 proteins to provide a platform for the specialized polymerases to bypass the lesion and resume DNA replication. After translesion synthesis, extension is required. This extension can be carried out by a replicative polymerase if the TLS is error-free, as in the case of Pol η, yet if TLS results in a mismatch, a specialized polymerase is needed to extend it; Pol ζ. Pol ζ is unique in that it can extend terminal mismatches, whereas more processive polymerases cannot. So when a lesion is encountered, the replication fork will stall, PCNA will switch from a processive polymerase to a TLS polymerase such as Pol ι to fix the lesion, then PCNA may switch to Pol ζ to extend the mismatch, and last PCNA will switch to the processive polymerase to continue replication.

Global Response to DNA Damage

Cells exposed to ionizing radiation, ultraviolet light or chemicals are prone to acquire multiple sites of bulky DNA lesions and double-strand breaks. Moreover, DNA damaging agents can damage other biomolecules such as proteins, carbohydrates, lipids, and RNA. The accumulation of damage, to be specific, double-strand breaks or adducts stalling the replication forks, are among known stimulation signals for a global response to DNA damage. The global response to damage is an act directed toward the cells' own preservation and triggers multiple pathways of macromolecular repair, lesion bypass, tolerance, or apoptosis. The common features of global response are induction of multiple genes, cell cycle arrest, and inhibition of cell division.

Initial Steps

The packaging of eukaryotic DNA into chromatin presents a barrier to all DNA-based processes that require recruitment of enzymes to their sites of action. To allow DNA repair, the chromatin must be remodeled. In eukaryotes, ATP dependent chromatin remodeling complexes and histone-modifying enzymes are two predominant factors employed to accomplish this remodeling process.

Chromatin relaxation occurs rapidly at the site of a DNA damage. In one of the earliest steps, the stress-activated protein kinase, c-Jun N-terminal kinase (JNK), phosphorylates SIRT6 on serine 10 in response to double-strand breaks or other DNA damage. This post-translational modification facilitates the mobilization of SIRT6 to DNA damage sites, and is required for efficient recruitment of poly (ADP-ribose) polymerase 1 (PARP1) to DNA break sites and for efficient repair of DSBs. PARP1 protein starts to appear at DNA damage sites in less than a second, with half maximum accumulation within 1.6 seconds after the damage occurs. PARP1 synthesizes polymeric adenosine diphosphate ribose (poly (ADP-ribose) or PAR) chains on itself. Next the chromatin remodeler ALC1 quickly attaches to the product of PARP1 action, a poly-ADP ribose chain, and ALC1 completes arrival at the DNA damage within 10 seconds of the occurrence of the damage. About half of the maximum chromatin relaxation, presumably due to action of ALC1, occurs by 10 seconds. This then allows recruitment of the DNA repair enzyme MRE11, to initiate DNA repair, within 13 seconds.

γH2AX, the phosphorylated form of H2AX is also involved in the early steps leading to chromatin decondensation after DNA double-strand breaks. The histone variant H2AX constitutes about 10% of the H2A histones in human chromatin. γH2AX (H2AX phosphorylated on serine 139) can be detected as soon as 20 seconds after irradiation of cells (with DNA double-strand break formation), and half maximum accumulation of γH2AX occurs in one minute. The extent of chromatin with phosphorylated γH2AX is about two million base pairs at the site of a DNA double-strand break. γH2AX does not, itself, cause chromatin decondensation, but within 30 seconds of irradiation, RNF8 protein can be detected in association with γH2AX. RNF8 mediates extensive chromatin decondensation, through its subsequent interaction with CHD4, a component of the nucleosome remodeling and deacetylase complex NuRD.

DDB2 occurs in a heterodimeric complex with DDB1. This complex further complexes with the ubiquitin ligase protein CUL4A and with PARP1. This larger complex rapidly associates with UV-induced damage within chromatin, with half-maximum association completed in 40 seconds. The PARP1 protein, attached to both DDB1 and DDB2, then PARylates (creates a poly-ADP ribose chain) on DDB2 that attracts the DNA remodeling protein ALC1. Action of ALC1 relaxes the chromatin at the site of UV damage to DNA. This relaxation allows other proteins in the nucleotide excision repair pathway to enter the chromatin and repair UV-induced cyclobutane pyrimidine dimer damages.

After rapid chromatin remodeling, cell cycle checkpoints are activated to allow DNA repair to occur before the cell cycle progresses. First, two kinases, ATM and ATR are activated within 5 or 6 minutes after DNA is damaged. This is followed by phosphorylation of the cell cycle checkpoint protein Chk1, initiating its function, about 10 minutes after DNA is damaged.

DNA Damage Checkpoints

After DNA damage, cell cycle checkpoints are activated. Checkpoint activation pauses the cell cycle and gives the cell time to repair the damage before continuing to divide. DNA damage checkpoints occur at the G1/S and G2/M boundaries. An intra-S checkpoint also exists. Checkpoint activation is controlled by two master kinases, ATM and ATR. ATM responds to DNA double-strand breaks and disruptions in chromatin structure, whereas ATR primarily responds to stalled replication forks. These kinases phosphorylate downstream targets in a signal transduction cascade, eventually leading to cell cycle arrest. A class of checkpoint mediator proteins including BRCA1, MDC1, and 53BP1 has also been identified. These proteins seem to be required for transmitting the checkpoint activation signal to downstream proteins.

DNA damage checkpoint is a signal transduction pathway that blocks cell cycle progression in G1, G2 and metaphase and slows down the rate of S phase progression when DNA is damaged. It leads to a pause in cell cycle allowing the cell time to repair the damage before continuing to divide.

Checkpoint Proteins can be separated into four groups: phosphatidylinositol 3-kinase (PI3K)-like protein kinase, proliferating cell nuclear antigen (PCNA)-like group, two serine/threonine(S/T) kinases and their adaptors. Central to all DNA damage induced checkpoints responses is a pair of large protein kinases belonging to the first group of PI3K-like protein kinases-the ATM (Ataxia telangiectasia mutated) and ATR (Ataxia- and Rad-related) kinases, whose sequence and functions have been well conserved in evolution. All DNA damage response requires either ATM or ATR because they have the ability to bind to the chromosomes at the site of DNA damage, together with accessory proteins that are platforms on which DNA damage response components and DNA repair complexes can be assembled.

An important downstream target of ATM and ATR is p53, as it is required for inducing apoptosis following DNA damage. The cyclin-dependent kinase inhibitor p21 is induced by both p53-dependent and p53-independent mechanisms and can arrest the cell cycle at the G1/S and G2/M checkpoints by deactivating cyclin/cyclin-dependent kinase complexes.

The Prokaryotic SOS Response

The SOS response is the changes in gene expression in *Escherichia coli* and other bacteria in response to extensive DNA damage. The prokaryotic SOS system is regulated by two key proteins: LexA and RecA. The LexA homodimer is a transcriptional repressor that binds to operator sequences commonly referred to as SOS boxes. In *Escherichia coli* it is known that LexA regulates transcription of approximately 48 genes including the lexA and recA genes. The SOS response is known to be widespread in the Bacteria domain, but it is mostly absent in some bacterial phyla, like the Spirochetes. The most common cellular signals activating the SOS response are regions of single-stranded DNA (ssDNA), arising from stalled replication forks or double-strand breaks, which are processed by DNA helicase to separate the two DNA strands. In the initiation step, RecA protein binds to ssDNA in an ATP hydrolysis driven reaction creating RecA–ssDNA filaments. RecA–ssDNA filaments activate LexA autoprotease activity, which ultimately leads to cleavage of LexA dimer and subsequent LexA degradation. The loss of LexA repressor induces transcription of the SOS genes and allows for further signal induction, inhibition of cell division and an increase in levels of proteins responsible for damage processing.

In *Escherichia coli*, SOS boxes are 20-nucleotide long sequences near promoters with palindromic structure and a high degree of sequence conservation. In other classes and phyla, the sequence of SOS boxes varies considerably, with different length and composition, but it is always highly conserved and one of the strongest short signals in the genome. The high information content of SOS boxes permits differential binding of LexA to different promoters and allows for timing of the SOS response. The lesion repair genes are induced at the beginning of SOS response. The error-prone translesion polymerases, for example, UmuCD'2 (also called DNA polymerase V), are induced later on as a last resort. Once the DNA damage is repaired or bypassed using polymerases or through recombination, the amount of single-stranded DNA in cells is decreased, lowering the amounts of RecA filaments decreases cleavage activity of LexA homodimer, which then binds to the SOS boxes near promoters and restores normal gene expression.

Eukaryotic Transcriptional Responses to DNA Damage

Eukaryotic cells exposed to DNA damaging agents also activate important defensive pathways by inducing multiple proteins involved in DNA repair, cell cycle checkpoint control, protein trafficking and degradation. Such genome wide transcriptional response is very complex and tightly regulated, thus allowing coordinated global response to damage. Exposure of yeast *Saccharomyces cerevisiae* to DNA damaging agents results in overlapping but distinct transcriptional profiles. Similarities to environmental shock response indicates that a general global stress response pathway exist at the level of transcriptional activation. In contrast, different human cell types respond to damage differently indicating an absence of a common global response. The probable explanation for this difference between yeast and human cells may be in the heterogeneity of mammalian cells. In an animal different types of cells are distributed among different organs that have evolved different sensitivities to DNA damage.

In general global response to DNA damage involves expression of multiple genes responsible for postreplication repair, homologous recombination, nucleotide excision repair, DNA damage checkpoint, global transcriptional activation, genes controlling mRNA decay, and many others. A large amount of damage to a cell leaves it with an important decision: undergo apoptosis and die, or survive at the cost of living with a modified genome. An increase in tolerance to damage can lead to an increased rate of survival that will allow a greater accumulation of mutations. Yeast Rev1 and human polymerase η are members of Y family translesion DNA polymerases present during global response to DNA damage and are responsible for enhanced mutagenesis during a global response to DNA damage in eukaryotes.

Free Radical Damage to DNA

Free radical damage to DNA can occur as a result of exposure to ionizing radiation or to radiomimetic compounds. Damage to DNA as a result of free radical attack is called indirect DNA damage because the radicals formed can diffuse throughout the body and affect other organs. Malignant melanoma can be caused by indirect DNA damage because it is found in parts of the body not exposed to sunlight. DNA is vulnerable to radical attack because of the very labile hydrogens that can be abstracted and the prevalence of double bonds in the DNA bases that radicals can easily add to.

Damage via Radiation Exposure

Radiolysis of intracellular water by ionizing radiation creates peroxides, which are relatively stable precursors to hydroxyl radicals. 60%- 70% of cellular DNA damage is caused by hydroxyl radicals, yet hydroxyl radicals are so reactive that they can only diffuse one or two molecular diameters before reacting with cellular components. Thus, hydroxyl radicals must be formed immediately adjacent to nucleic acids in order to react. Radiolysis of water creates peroxides that can act as diffusable, latent forms of hydroxyl radicals. Some metal ions in the vicinity of DNA generate the hydroxyl radicals from peroxide.

$$H_2O + hv \rightarrow H_2O^+ + e^-$$

$$H_2O + e^- \rightarrow H_2O^-$$

$$H_2O^+ \rightarrow H^+ + OH^\cdot$$

$$H_2O^- \rightarrow OH^- + H^\cdot$$

$$2\,OH^\cdot \rightarrow H_2O_2$$

Free radical damage to DNA is thought to result in mutations that may lead to some cancers.

The Fenton Reaction

The Fenton reaction results in the creation of hydroxyl radicals from hydrogen peroxide and an Iron (II) catalyst. Iron(III) is regenerated via the Haber-Weiss reaction. Transition metals with a free coordination site are capable of reducing peroxides to hydroxyl radicals. Iron is believed to be the metal responsible for the creation of hydroxyl radicals because it exists at the highest concentration of any transition metal in most living organisms. The Fenton reaction is possible because transition metals can exist in more than one oxidation state and their valence electrons may be unpaired, allowing them to participate in one-electron redox reactions.

$$Fe2+ + H2O2 \rightarrow Fe3+ + OH\cdot + OH-$$

The creation of hydroxyl radicals by iron(II) catalysis is important because iron(II) can be found coordinated with, and therefore in close proximity to, DNA. This reaction allows for hydrogen peroxide created by radiolysis of water to diffuse to the nucleus and react with Iron (II) to produce hydroxyl radicals, which in turn react with DNA. The location and binding of Iron (II) to DNA may play an important role in determining the substrate and nature of the radical attack on the DNA. The Fenton reaction generates two types of oxidants, Type I and Type II. Type I oxidants are moderately sensitive to peroxides and ethanol. Type I and Type II oxidants preferentially cleave at the specific sequences.

Radical Hydroxyl Attack

Hydroxyl radicals can attack the deoxyribose DNA backbone and bases, potentially causing a plethora of lesions that can be cytotoxic or mutagenic. Cells have developed complex and efficient repair mechanisms to fix the lesions. In the case of free radical attack on DNA, base-excision repair is the repair mechanism used. Hydroxyl radical reactions with the deoxyribose sugar backbone are initiated by hydrogen abstraction from a deoxyribose carbon, and the predominant consequence is eventual strand breakage and base release. The hydroxyl radical reacts with the various hydrogen atoms of the deoxyribose in the order 5′ H > 4′ H > 3′ H ≈ 2′ H ≈ 1′ H. This order of reactivity parallels the exposure to solvent of the deoxyribose hydrogens.

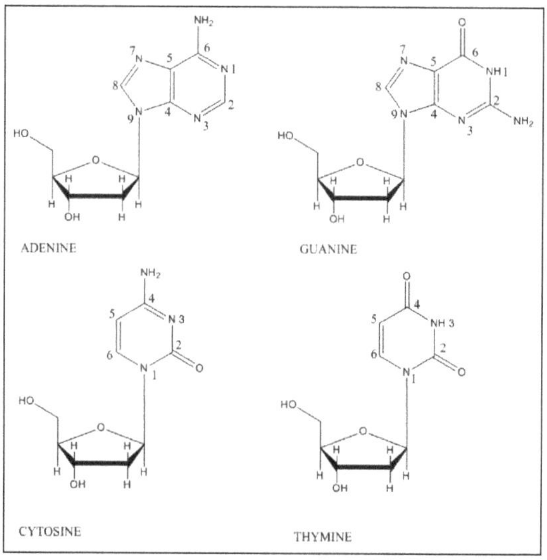

Radical hydroxyl attacks can form baseless sites.

Hydroxyl radicals react with DNA bases via addition to the electron-rich, pi bonds. These pi bonds in the bases are located between C5-C6 of pyrimidines and N7-C8 in purines. Upon addition of the hydroxyl radical, many stable products can be formed. In general, radical hydroxyl attacks on base moieties do not cause altered sugars or strand breaks except when the modifications labilize the N-glycosyl bond, allowing the formation of baseless sites that are subject to beta-elimination.

Abasic Sites

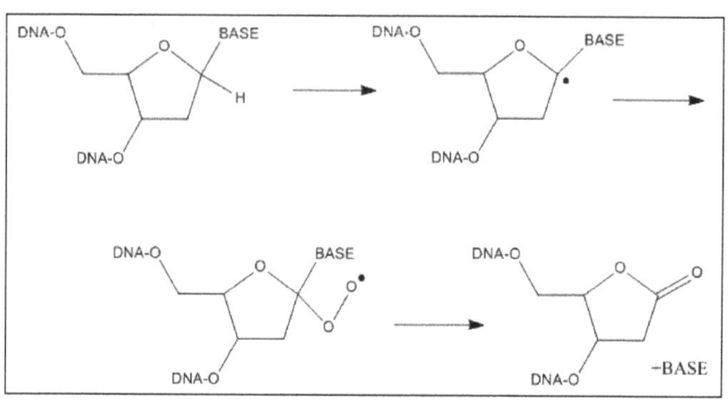

Route of deoxyribonolactone formation.

Hydrogen abstraction from the 1'-deoxyribose carbon by the hydroxyl radical creates a 1 '-deoxyribosyl radical. The radical can then react with molecular oxygen, creating a peroxyl radical which can be reduced and dehydrated to yield a 2'-deoxyribonolactone and free base. A deoxyribonolactone is mutagenic and resistant to repair enzymes. Thus, an abasic site is created.

Radical Damage through Radiomimetic Compounds

Radical damage to DNA can also occur through the interaction of DNA with certain natural products known as radiomimetic compounds, molecular compounds which affect DNA in similar ways to radiation exposure. Radiomimetic compounds induce double-strand breaks in DNA via highly specific, concerted free-radical attacks on the deoxyribose moieties in both strands of DNA.

General Mechanism

Many radiomimetic compounds are enediynes, which undergo the Bergman cyclization reaction to produce a 1,4-didehydrobenzene diradical. The 1,4-didehydrobenzene diradical is highly reactive, and will abstract hydrogens from any possible hydrogen-donor.

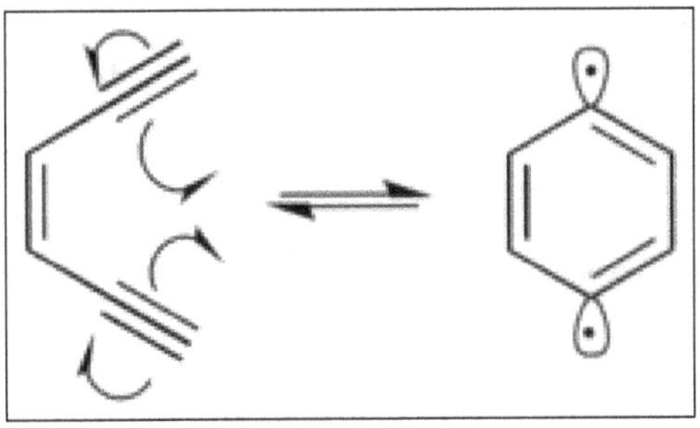

In the presence of DNA, the 1,4-didehydrobenzene diradical abstracts hydrogens from the deoxyribose sugar backbone, predominantly at the C-1', C-4' and C-5' positions. Hydrogen abstraction causes radical formation at the reacted carbon. The carbon radical reacts with molecular oxygen, which leads to a strand break in the DNA through a variety of mechanisms. 1,4-Didehydrobenzene is able to position itself in such a way that it can abstract proximal hydrogens from both strands of DNA. This produces a double-strand break in the DNA, which can lead to cellular apoptosis if not repaired.

Enediynes generally undergo the Bergman cyclization at temperatures exceeding 200 °C. However, incorporating the enediyne into a 10-membered cyclic hydrocarbon makes the reaction more thermodynamically favorable by releasing the ring strain of the reactants. This allows for the Bergman cyclization to occur at 37 °C, the biological temperature of humans. Molecules which incorporate enediynes into these larger ring structures have been found to be extremely cytotoxic.

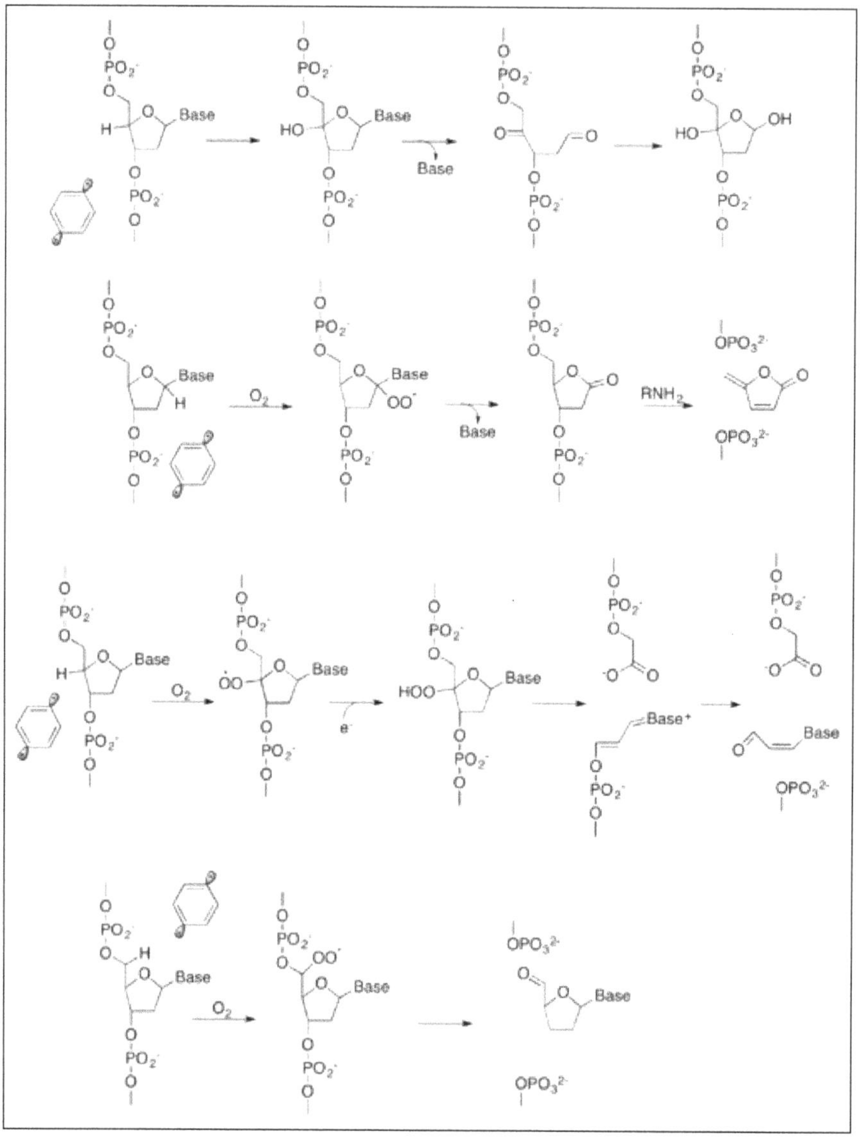

Natural Products

Enediynes are present in many complicated natural products. They were originally discovered in the early 1980s during a search for new anticancer products produced by microorganisms. Calicheamicin was one of the first such products identified and was originally found in a soil sample taken from Kerrville, Texas. These compounds are synthesized by bacteria as defense mechanisms due to their ability to cleave DNA through the formation of 1,4-didehydrobenzene from the enediyne component of the molecule.

Calicheamicin and other related compounds share several common characteristics. The extended structures attached to the enediyne allow the compound to specifically bind DNA, in most cases to the minor groove of the double helix. Additionally, part of the molecule is known as the "trigger" which, under specific physiological conditions, activates the enediyne, known as the "warhead" and 1,4-didehydrobenzene is generated.

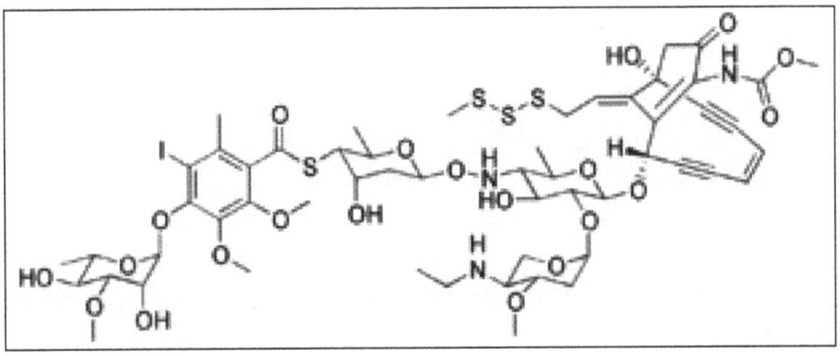

Three classes of enediynes have since been identified: calicheamicin, dynemicin, and chromoprotein based products.

The calicheamicin types are defined by a methyl trisulfide group that is involved in triggering the molecule by the following mechanism.

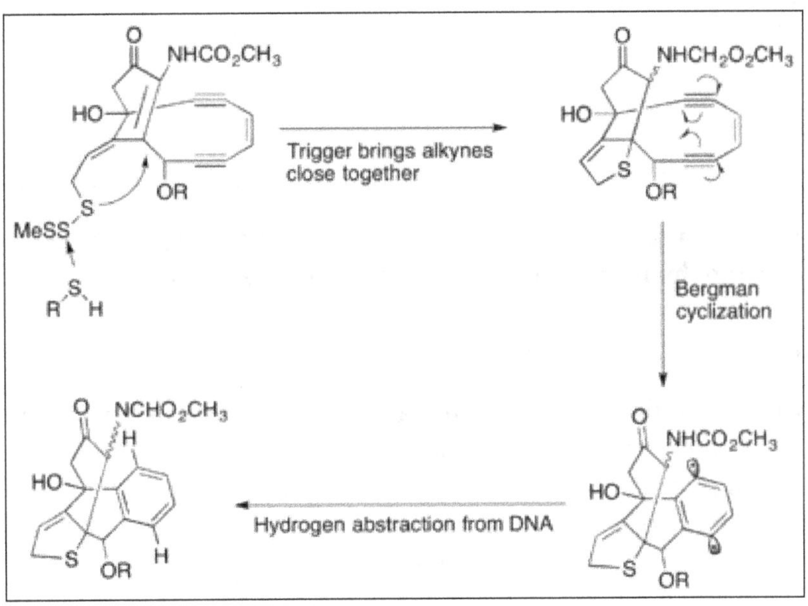

Calicheamicin and the closely related esperamicin have been used as anticancer drugs due to their high toxicity and specificity.

Dynemicin and its relatives are characterized by the presence of an anthraquinone and enediyne core. The anthraquinone component allows for specific binding of DNA at the 3' side of purine bases through intercalation, a site that is different from calicheamicin. Its ability to cleave DNA is greatly increased in the presence of NADPH and thiol compounds. This compound has also found prominence as an antitumor agent.

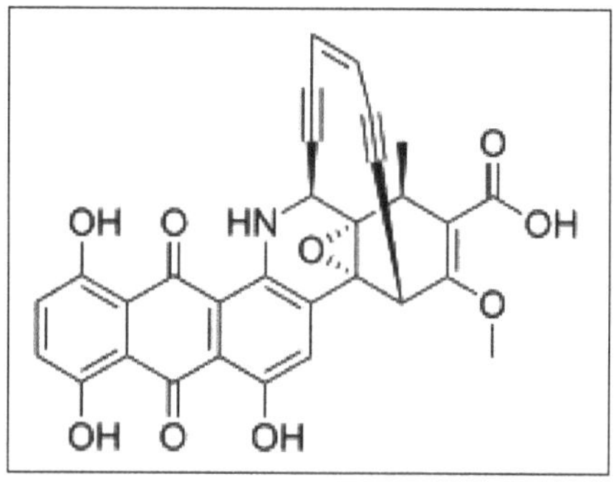

Chromoprotein enediynes are characterized by an unstable chromophore enediyne bound to an apoprotein.

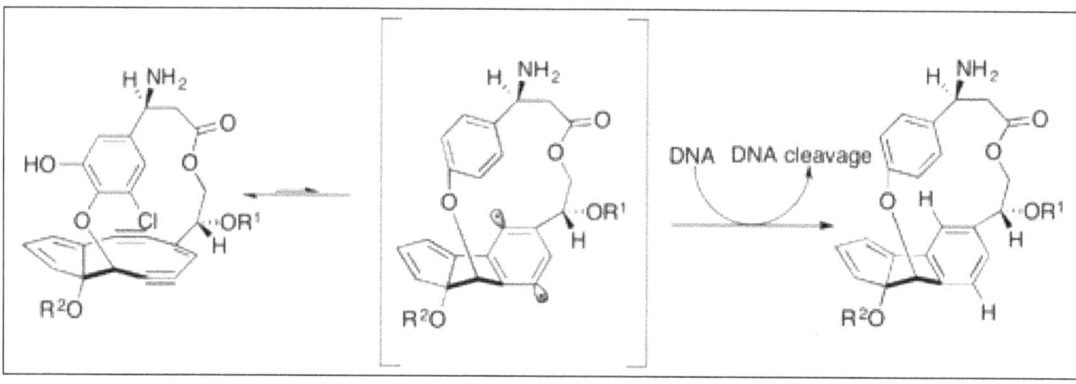

The chromophore is unreactive when bound to the apoprotein. Upon its release, it reacts to form 1,4-didehydrobenzene and subsequently cleaves DNA.

Antitumor Ability

Most enediynes, including the ones listed above, have been used as potent antitumor antibiotics due to their ability to efficiently cleave DNA. Calicheamicin and esperamicin are the two most commonly used types due to their high specificity when binding to DNA, which minimizes unfavorable side reactions. They have been shown to be especially useful for treating acute myeloid leukemia.

Additionally, calicheamicin is able to cleave DNA at low concentrations, proving to be up to 1000 times more effective than adriamycin at combating certain types of tumors. In all cases, cells lack the ability to repair double-stranded DNA breaks, making these compounds especially effective for treating tumor cells.

The free radical mechanism to treat certain types of cancers extends beyond enediynes. Tirapazamine generates a free radical under anoxic conditions instead of the trigger mechanism of an enediyne. The free radical then continues on to cleave DNA in a similar manner to 1,4-didehydrobenzene in order to treat cancerous cells. It is currently in Phase III trials.

DNA Repair Modulation

Hereditary DNA Repair Disorders

Defects in the NER mechanism are responsible for several genetic disorders, including:

- Xeroderma pigmentosum: Hypersensitivity to sunlight/UV, resulting in in-creased skin cancer incidence and premature aging.
- Cockayne syndrome: Hypersensitivity to UV and chemical agents.
- Trichothiodystrophy: Sensitive skin, brittle hair and nails.

Mental retardation often accompanies the latter two disorders, suggesting increased vulnerability of developmental neurons.

Other DNA repair disorders include:

- Werner's syndrome: Premature aging and retarded growth.
- Bloom's syndrome: Sunlight hypersensitivity, high incidence of malignancies (especially leukemias).
- Ataxia telangiectasia: Sensitivity to ionizing radiation and some chemical agents.

All of the above diseases are often called "segmental progerias" ("accelerated aging diseases") because their victims appear elderly and suffer from aging-related diseases at an abnormally young age, while not manifesting all the symptoms of old age.

Other diseases associated with reduced DNA repair function include Fanconi anemia, hereditary breast cancer and hereditary colon cancer.

Cancer

Because of inherent limitations in the DNA repair mechanisms, if humans lived long enough, they would all eventually develop cancer. There are at least 34 Inherited human DNA repair gene mutations that increase cancer risk. Many of these mutations cause DNA repair to be less effective than normal. In particular, Hereditary nonpolyposis colorectal cancer (HNPCC) is strongly associated with specific mutations in the DNA mismatch repair pathway. *BRCA1* and *BRCA2*, two important genes whose mutations confer a hugely increased risk of breast cancer on carriers, are both associated with a large number of DNA repair pathways, especially NHEJ and homologous recombination.

Cancer therapy procedures such as chemotherapy and radiotherapy work by overwhelming the capacity of the cell to repair DNA damage, resulting in cell death. Cells that are most rapidly dividing – most typically cancer cells – are preferentially affected. The side-effect is that other non-cancerous but rapidly dividing cells such as progenitor cells in the gut, skin, and hematopoietic system are also affected. Modern cancer treatments attempt to localize the DNA damage to cells and tissues only associated with cancer, either by physical means (concentrating the therapeutic agent in the region of the tumor) or by biochemical means (exploiting a feature unique to cancer cells in the body). In the context of therapies targeting DNA damage response genes, the latter approach has been termed 'synthetic lethality'.

Perhaps the most well-known of these 'synthetic lethality' drugs is the poly(ADP-ribose) polymerase 1 (PARP1) inhibitor olaparib, which was approved by the Food and Drug Administration in 2015 for the treatment in women of BRCA-defective ovarian cancer. Tumor cells with partial loss of DNA damage response (specifically, homologous recombination repair) are dependent on another mechanism – single-strand break repair – which is a mechanism consisting, in part, of the PARP1 gene product. Olaparib is combined with chemotherapeutics to inhibit single-strand break repair induced by DNA damage caused by the co-administered chemotherapy. Tumor cells relying on this residual DNA repair mechanism are unable to repair the damage and hence are not able to survive and proliferate, whereas normal cells can repair the damage with the functioning homologous recombination mechanism.

Many other drugs for use against other residual DNA repair mechanisms commonly found in cancer are currently under investigation. However, synthetic lethality therapeutic approaches have been questioned due to emerging evidence of acquired resistance, achieved through rewiring of DNA damage response pathways and reversion of previously-inhibited defects.

DNA Repair Defects in Cancer

It has become apparent over the past several years that the DNA damage response acts as a barrier to the malignant transformation of preneoplastic cells. Previous studies have shown an elevated DNA damage response in cell-culture models with oncogene activation and preneoplastic colon adenomas. DNA damage response mechanisms trigger cell-cycle arrest, and attempt to repair DNA lesions or promote cell death/senescence if repair is not possible. Replication stress is observed in preneoplastic cells due to increased proliferation signals from oncogenic mutations. Replication stress is characterized by: increased replication initiation/origin firing; increased transcription and collisions of transcription-replication complexes; nucleotide deficiency; increase in reactive oxygen species (ROS).

Replication stress, along with the selection for inactivating mutations in DNA damage response genes in the evolution of the tumor, leads to downregulation and loss of some DNA damage response mechanisms, and hence loss of DNA repair and senescence/programmed cell death. In experimental mouse models, loss of DNA damage response-mediated cell senescence was observed after using a short hairpin RNA (shRNA) to inhibit the double-strand break response kinase ataxia telangiectasia (ATM), leading to increased tumor size and invasiveness. Humans born with inherited defects in DNA repair mechanisms (for example, Li-Fraumeni syndrome) have a higher cancer risk.

The prevalence of DNA damage response mutations differs across cancer types; for example, 30% of breast invasive carcinomas have mutations in genes involved in homologous recombination. In cancer, downregulation is observed across all DNA damage response mechanisms (base excision repair (BER), nucleotide excision repair (NER), DNA mismatch repair (MMR), homologous recombination repair (HR), non-homologous end joining (NHEJ) and translesion DNA synthesis (TLS). As well as mutations to DNA damage repair genes, mutations also arise in the genes responsible for arresting the cell cycle to allow sufficient time for DNA repair to occur, and some genes are involved in both DNA damage repair and cell cycle checkpoint control, for example ATM and checkpoint kinase 2 (CHEK2) – a tumor suppressor that is often absent or downregulated in non-small cell lung cancer.

Table: Genes involved in DNA damage response pathways and frequently mutated in cancer (HR = homologous recombination; NHEJ = non-homologous end joining; SSA = single-strand annealing; FA = fanconi anemia pathway; BER = base excision repair; NER = nucleotide excision repair; MMR = mismatch repair).

	HR	NHEJ	SSA	FA	BER	NER	MMR
ATM	x	x	x				
ATR	x	x	x				
PAXIP	x	x					
RPA	x		x			x	
BRCA1	x			x			
BRCA2	x			x			
RAD51	x			x			
RFC	x				x	x	
XRCC1					x	x	
PCNA					x	x	x
PARP1		x			x		
ERCC1	x		x	x		x	
MSH3	x		x				x

Epigenetic DNA Repair Defects in Cancer

Classically, cancer has been viewed as a set of diseases that are driven by progressive genetic abnormalities that include mutations in tumour-suppressor genes and oncogenes, and chromosomal aberrations. However, it has become apparent that cancer is also driven by epigenetic alterations.

Epigenetic alterations refer to functionally relevant modifications to the genome that do not involve a change in the nucleotide sequence. Examples of such modifications are changes in DNA methylation (hypermethylation and hypomethylation) and histone modification, changes in chromosomal architecture (caused by inappropriate expression of proteins such as HMGA2 or HMGA1) and changes caused by microRNAs. Each of these epigenetic alterations serves to regulate gene expression without altering the underlying DNA sequence. These changes usually remain through cell divisions, last for multiple cell generations, and can be considered to be epimutations (equivalent to mutations).

While large numbers of epigenetic alterations are found in cancers, the epigenetic alterations in DNA repair genes, causing reduced expression of DNA repair proteins, appear to be particularly important. Such alterations are thought to occur early in progression to cancer and to be a likely cause of the genetic instability characteristic of cancers.

Reduced expression of DNA repair genes causes deficient DNA repair. When DNA repair is deficient DNA damages remain in cells at a higher than usual level and these excess damages cause increased frequencies of mutation or epimutation. Mutation rates increase substantially in cells defective in DNA mismatch repair or in homologous recombinational repair (HRR). Chromosomal rearrangements and aneuploidy also increase in HRR defective cells.

Higher levels of DNA damage not only cause increased mutation, but also cause increased epimutation. During repair of DNA double strand breaks, or repair of other DNA damages, incompletely cleared sites of repair can cause epigenetic gene silencing.

Deficient expression of DNA repair proteins due to an inherited mutation can cause increased risk of cancer. Individuals with an inherited impairment in any of 34 DNA repair genes have an increased risk of cancer, with some defects causing up to a 100% lifetime chance of cancer (e.g. p53 mutations). However, such germline mutations (which cause highly penetrant cancer syndromes) are the cause of only about 1 percent of cancers.

Frequencies of Epimutations in DNA Repair Genes

A chart of common DNA damaging agents, examples of lesions they cause in DNA, and pathways used to repair these lesions. Also shown are many of the genes in these pathways, an indication of which genes are epigenetically regulated to have reduced (or increased) expression in various cancers. It also shows genes in the error prone microhomology-mediated end joining pathway with increased expression in various cancers.

Deficiencies in DNA repair enzymes are occasionally caused by a newly arising somatic mutation in a DNA repair gene, but are much more frequently caused by epigenetic alterations that reduce or silence expression of DNA repair genes. For example, when 113 colorectal cancers were examined in sequence, only four had a missense mutation in the DNA repair gene MGMT, while the majority had reduced MGMT expression due to methylation of the MGMT promoter region (an epigenetic alteration). Five different studies found that between 40% and 90% of colorectal cancers have reduced MGMT expression due to methylation of the MGMT promoter region.

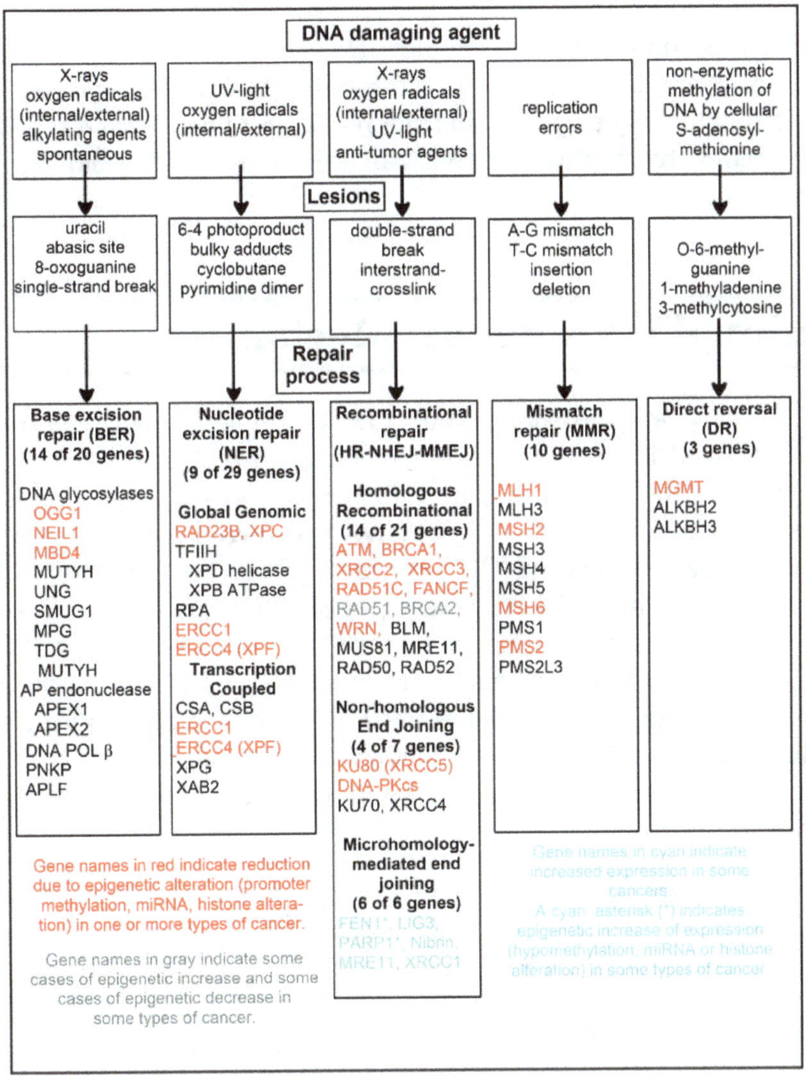

Similarly, out of 119 cases of mismatch repair-deficient colorectal cancers that lacked DNA repair gene PMS2 expression, PMS2 was deficient in 6 due to mutations in the PMS2 gene, while in 103 cases PMS2 expression was deficient because its pairing partner MLH1 was repressed due to promoter methylation (PMS2 protein is unstable in the absence of MLH1). In the other 10 cases, loss of PMS2 expression was likely due to epigenetic overexpression of the microRNA, miR-155, which down-regulates MLH1.

In further examples, epigenetic defects were found at frequencies of between 13%–100% for the DNA repair genes BRCA1, WRN, FANCB, FANCF, MGMT, MLH1, MSH2, MSH4, ERCC1, XPF, NEIL1 and ATM. These epigenetic defects occurred in various cancers (e.g. breast, ovarian, colorectal and head and neck). Two or three deficiencies in the expression of ERCC1, XPF or PMS2 occur simultaneously in the majority of the 49 colon cancers evaluated by Facista et al.

The chart in this section shows some frequent DNA damaging agents, examples of DNA lesions they cause, and the pathways that deal with these DNA damages. At least 169 enzymes are either directly employed in DNA repair or influence DNA repair processes. Of these, 83 are directly employed in repairing the 5 types of DNA damages illustrated in the chart.

Additional DNA Repair Mechanisms

NER and photoreactivation are not the only methods of DNA repair. For instance, base excision repair (BER) is the predominant mechanism that handles the spontaneous DNA damage caused by free radicals and other reactive species generated by metabolism. Bases can become oxidized, alkylated, or hydrolyzed through interactions with these agents. For example, methyl (CH_3) chemical groups are frequently added to guanine to form 7-methylguanine; alternatively, purine groups may be lost. All such changes result in abnormal bases that must be removed and replaced. Thus, enzymes known as DNA glycosylases remove damaged bases by literally cutting them out of the DNA strand through cleavage of the covalent bonds between the bases and the sugar-phosphate backbone. The resulting gap is then filled by a specialized repair polymerase and sealed by ligase. Many such enzymes are found in cells, and each is specific to certain types of base alterations.

Yet another form of DNA damage is double-strand breaks, which are caused by ionizing radiation, including gamma rays and X-rays. These breaks are highly deleterious. In addition to interfering with transcription or replication, they can lead to chromosomal rearrangements, in which pieces of one chromosome become attached to another chromosome. Genes are disrupted in this process, leading to hybrid proteins or inappropriate activation of genes. A number of cancers are associated with such rearrangements. Double-strand breaks are repaired through one of two mechanisms: nonhomologous end joining (NHEJ) or homologous recombination repair (HRR). In NHEJ, an enzyme called DNA ligase IV uses overhanging pieces of DNA adjacent to the break to join and fill in the ends. Additional errors can be introduced during this process, which is the case if a cell has not completely replicated its DNA in preparation for division. In contrast, during HRR, the homologous chromosome itself is used as a template for repair.

Mutations in an organism's DNA are a part of life. Our genetic code is exposed to a variety of insults that threaten its integrity. But, a rigorous system of checks and balances is in place through the DNA repair machinery. The errors that slip through the cracks may sometimes be associated with disease, but they are also a source of variation that is acted upon by longer-term processes, such as evolution and natural selection.

References

- DNA-Damage-Causes-Effects: cellbiolabs.com, Retrieved 8 January, 2020
- Köhler, Kerstin; Ferreira, Pedro; Pfander, Boris; Boos, Dominik (2016). The Initiation of DNA Replication in Eukaryotes. Springer, Cham. Pp. 443–460. Doi:10.1007/978-3-319-24696-3-22. ISBN 9783319246949
- Nature Reviews Series: DNA damage". Nature Reviews Molecular Cell Biology. 5 July 2017. Retrieved 7 November 2018
- De Bont R, van Larebeke N (2004). "Endogenous DNA damage in humans: a review of quantitative data". Mutagenesis. 19 (3): 169–185. Doi:10.1093/mutage/geh025. PMID 15123782
- Wei, Qingyi; Lei Li; David Chen (2007). DNA Repair, Genetic Instability, and Cancer. World Scientific. ISBN 978-981-270-014-8
- Lynch, MD. (2006). "How does cellular senescence prevent cancer?". DNA Cell Biol. 25 (2): 69–78. Doi:10.1089/dna.2006.25.69. PMID 16460230
- Broad, William J. (7 October 2015). "Nobel Prize in Chemistry Awarded to Tomas Lindahl, Paul Modrich and Aziz Sancar for DNA Studies". The New York Times. Retrieved 7 October 2015
- Xinrong Chen and Tao Chen (2011). "Chapter 18: Roles of microrna in DNA Damage and Repair", DNA Repair, Inna Kruman (ed.), ISBN 9789533076973, intech
- Dna-damage-repair-mechanisms-for-maintaining-dna, topicpage, scitable: nature.com, Retrieved 13 May, 2020

5

DNA Profiling and Sequencing

The process of determining the DNA characteristics of an individual is referred to as DNA profiling. DNA sequencing refers to the process of determining the nucleic acid sequencing in DNA. All the diverse aspects of DNA profiling and sequencing including their applications have been carefully analyzed in this chapter.

DNA Profiling

DNA profiling is the process where a specific DNA pattern, called a profile, is obtained from a person or sample of bodily tissue.

Even though we are all unique, most of our DNA is actually identical to other people's DNA. However, specific regions vary highly between people. These regions are called polymorphic. Differences in these variable regions between people are known as polymorphisms. Each of us inherits a unique combination of polymorphisms from our parents. DNA polymorphisms can be analysed to give a DNA profile.

Profiling Process

DNA Extraction

DNA isolation is a process of purification of DNA from sample using a combination of physical and chemical methods. The first isolation of DNA was done in 1869 by Friedrich Miescher. Currently it is a routine procedure in molecular biology or forensic analyses. For the chemical method, there are many different kits used for extraction, and selecting the correct one will save time on kit optimization and extraction procedures. PCR sensitivity detection is considered to show the variation between the commercial kits.

Basic Procedure

There are three basic and two optional steps in a DNA extraction.

- Cells which are to be studied need to be collected.
- Breaking the cell membranes open to expose the DNA along with the cytoplasm within (cell lysis).
    - Lipids from the cell membrane and the nucleus are broken down with detergents and surfactants.
    - Breaking proteins by adding a protease (optional).
    - Breaking RNA by adding an RNase (optional).
- The solution is treated with concentrated salt solution (saline) to make debris such as broken proteins, lipids and RNA to clump together.
- Centrifugation of the solution, which separates the clumped cellular debris from the DNA.
- DNA purification from detergents, proteins, salts and reagents used during cell lysis step. The most commonly used procedures are:
    - Ethanol precipitation usually by ice-cold ethanol or isopropanol. Since DNA is insoluble in these alcohols, it will aggregate together, giving a *pellet* upon centrifugation. Precipitation of DNA is improved by increasing of ionic strength, usually by adding sodium acetate.
    - Phenol–chloroform extraction in which phenol denatures proteins in the sample. After centrifugation of the sample, denaturated proteins stay in the organic phase while aqueous phase containing nucleic acid is mixed with the chloroform that removes phenol residues from solution.
    - Minicolumn purification that relies on the fact that the nucleic acids may bind (adsorption) to the solid phase (silica or other) depending on the pH and the salt concentration of the buffer.

Cellular and histone proteins bound to the DNA can be removed either by adding a protease or by having precipitated the proteins with sodium or ammonium acetate, or extracted them with a phenol-chloroform mixture prior to the DNA-precipitation.

After isolation, the DNA is dissolved in slightly alkaline buffer, usually in the TE buffer, or in ultra-pure water.

Method Selection

Some of the most common DNA extraction methods include organic extraction, Chelex extraction, and solid phase extraction. These methods consistently yield isolated DNA, but they differ in both the quality and the quantity of DNA yielded. When selecting a DNA extraction method, there are multiple factors to consider, including cost, time, safety, and risk of contamination.

Organic extraction involves the addition of and incubation in multiple different chemical solutions; including a lysis step, a phenol chloroform extraction, an ethanol precipitation, and washing steps. Organic extraction is often used in laboratories because it is cheap, and it yields large quantities of pure DNA. Though it is easy, there are many steps involved, and it takes longer than other methods. It also involves the unfavorable use of the toxic chemicals phenol and chloroform, and there is an increased risk of contamination due to transferring the DNA between multiple tubes. Several protocols based on organic extraction of DNA were effectively developed decades ago, though improved and more practical versions of these protocols have also been developed and published in the last years.

Chelex extraction method involves adding the Chelex resin to the sample, boiling the solution, then vortexing and centrifuging it. The cellular materials bind to the Chelex beads, while the DNA is available in the supernatant. The Chelex method is much faster and simpler than organic extraction, and it only requires one tube, which decreases the risk of DNA contamination. Unfortunately, Chelex extraction does not yield as much quantity and the DNA yielded is single-stranded, which means it can only be used for PCR-based analyses and not for RFLP.

Solid phase extraction such as using a spin-column based extraction method takes advantage of the fact that DNA binds to silica. The sample containing DNA is added to a column containing a silica gel or silica beads and chaotropic salts. The chaotropic salts disrupt the hydrogen bonding between strands and facilitate binding of the DNA to silica by causing the nucleic acids to become hydrophobic. This exposes the phosphate residues so they are available for adsorption. The DNA binds to the silica, while the rest of the solution is washed out using ehtanol to remove chaotropic salts and other unnecessary constituents. The DNA can then be rehydrated with aqueous low salt solutions allowing for elution of the DNA from the beads.

This method yields high-quality, largely double-stranded DNA which can be used for both PCR and RFLP analysis. This procedure can be automated and has a high throughput, although lower than the phenol-chloroform method. This is a one-step method i.e., the entire procedure is completed in one tube. This lowers the risk of contamination making it very useful for forensic extraction of DNA. Multiple solid phase extraction commercial kits are manufactured and marketed by different companies; the only problem is that they are more expensive than organic extraction or Chelex extraction.

Special Types

Specific techniques must be chosen for isolation of DNA from some samples. Typical samples with complicated DNA isolation are:

- Archaeological samples containing partially degraded DNA.

- Samples containing inhibitors of subsequent analysis procedures, most notably inhibitors of PCR, such as humic acid from soil, indigo and other fabric dyes or haemoglobin in blood.
- Samples from microorganisms with thick cellular wall, for example yeast.
- Samples containing mixed DNA from multiple sources.

Extrachromosomal DNA is generally easy to isolate, especially plasmids may be easily isolated by cell lysis followed by precipitation of proteins, which traps chromosomal DNA in insoluble fraction and after centrifugation, plasmid DNA can be purified from soluble fraction.

A Hirt DNA Extraction is an isolation of all extrachromosomal DNA in a mammalian cell. The Hirt extraction process gets rid of the high molecular weight nuclear DNA, leaving only low molecular weight mitochondrial DNA and any viral episomes present in the cell.

Detecting DNA

A diphenylamine (DPA) indicator will confirm the presence of DNA. This procedure involves chemical hydrolysis of DNA: when heated (e.g. ≥95 °C) in acid, the reaction requires a deoxyribose sugar and therefore is specific for DNA. Under these conditions, the 2-deoxyribose is converted to w-hydroxylevulinyl aldehyde, which reacts with the compound, diphenylamine, to produce a blue-colored compound. DNA concentration can be determined measuring the intensity of absorbance of the solution at the 600 nm with a spectrophotometer and comparing to a standard curve of known DNA concentrations.

Measuring the intensity of absorbance of the DNA solution at wavelengths 260 nm and 280 nm is used as a measure of DNA purity. DNA absorbs UV light at 260 and 280 nanometres, and aromatic proteins absorb UV light at 280 nm; a pure sample of DNA has a ratio of 1.8 at 260/280 and is relatively free from protein contamination. A DNA preparation that is contaminated with protein will have a 260/280 ratio lower than 1.8.

DNA can be quantified by cutting the DNA with a restriction enzyme, running it on an agarose gel, staining with ethidium bromide or a different stain and comparing the intensity of the DNA with a DNA marker of known concentration.

Using the Southern blot technique, this quantified DNA can be isolated and examined further using PCR and RFLP analysis. These procedures allow differentiation of the repeated sequences within the genome. It is these techniques which forensic scientists use for comparison, identification, and analysis.

Restriction Fragment Length Polymorphism

In molecular biology, restriction fragment length polymorphism (RFLP) is a technique that exploits variations in homologous DNA sequences, known as polymorphisms, in order to distinguish individuals, populations, or species or to pinpoint the locations of genes within a sequence. The term may refer to a polymorphism itself, as detected through the differing locations of restriction enzyme sites, or to a related laboratory technique by which such differences can be illustrated. In RFLP analysis, a DNA sample is digested into fragments by one or more restriction enzymes, and the resulting restriction fragments are then separated by gel electrophoresis according to their size.

Although now largely obsolete due to the emergence of inexpensive DNA sequencing technologies, RFLP analysis was the first DNA profiling technique inexpensive enough to see widespread application. RFLP analysis was an important early tool in genome mapping, localization of genes for genetic disorders, determination of risk for disease, and paternity testing.

RFLP Analysis

The basic technique for the detection of RFLPs involves fragmenting a sample of DNA with the application of a restriction enzyme, which can selectively cleave a DNA molecule wherever a short, specific sequence is recognized in a process known as a restriction digest. The DNA fragments produced by the digest are then separated by length through a process known as agarose gel electrophoresis and transferred to a membrane via the Southern blot procedure. Hybridization of the membrane to a labeled DNA probe then determines the length of the fragments which are complementary to the probe. A restriction fragment length polymorphism is said to occur when the length of a detected fragment varies between individuals, indicating non-identical sequence homologies. Each fragment length is considered an allele, whether it actually contains a coding region or not, and can be used in subsequent genetic analysis.

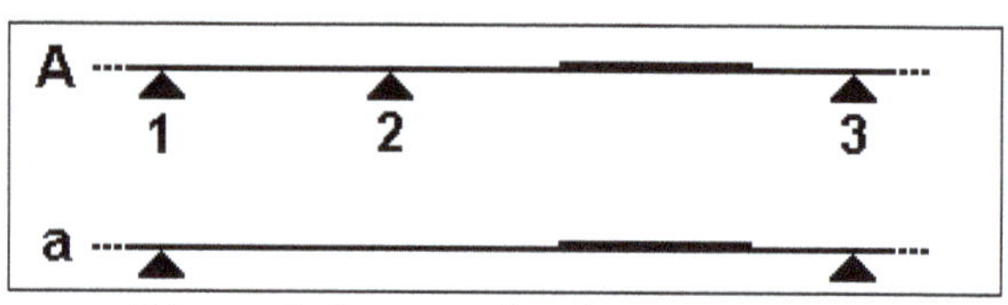

Schematic for RFLP by cleavage site loss.

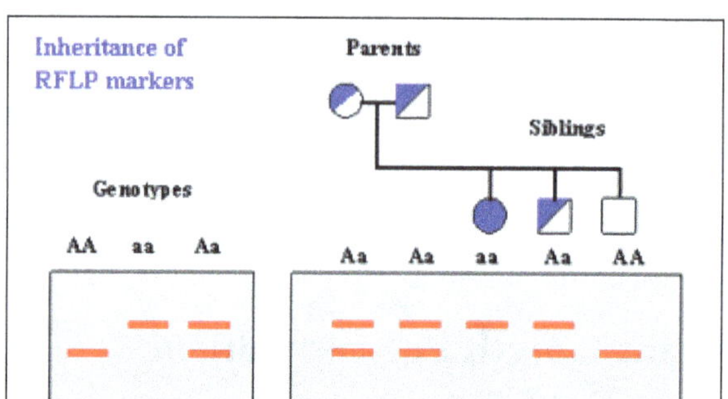

Analysis and inheritance of allelic RFLP fragments (NIH).

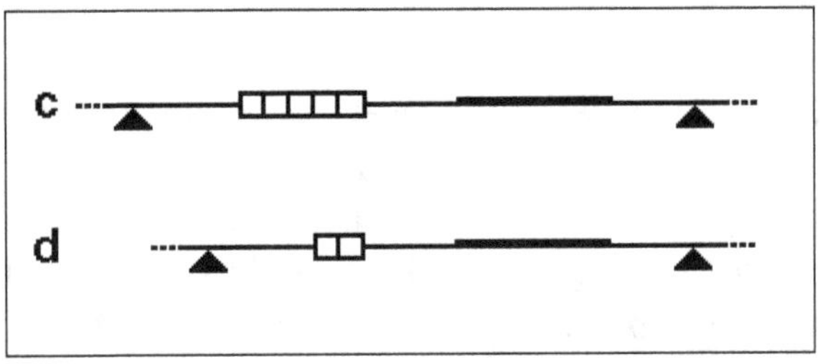

Schematic for RFLP by VNTR length variation.

Examples:

There are two common mechanisms by which the size of a particular restriction fragment can vary. In the first schematic, a small segment of the genome is being detected by a DNA probe (thicker line). In allele *A*, the genome is cleaved by a restriction enzyme at three nearby sites (triangles), but only the rightmost fragment will be detected by the probe. In allele *a*, restriction site 2 has been lost by a mutation, so the probe now detects the larger fused fragment running from sites 1 to 3. The second diagram shows how this fragment size variation would look on a Southern blot, and how each allele (two per individual) might be inherited in members of a family.

In the third schematic, the probe and restriction enzyme are chosen to detect a region of the genome that includes a variable number tandem repeat (VNTR) segment (boxes in schematic diagram). In allele *c*, there are five repeats in the VNTR, and the probe detects a longer fragment between the two restriction sites. In allele *d*, there are only two repeats in the VNTR, so the probe detects a shorter fragment between the same two restriction sites. Other genetic processes, such as insertions, deletions, translocations, and inversions, can also lead to polymorphisms. RFLP tests require much larger samples of DNA than do short tandem repeat (STR) tests.

Applications

Analysis of RFLP variation in genomes was formerly a vital tool in genome mapping and genetic disease analysis. If researchers were trying to initially determine the chromosomal location of a particular disease gene, they would analyze the DNA of members of a family afflicted by the disease, and look for RFLP alleles that show a similar pattern of inheritance as that of the disease. Once a disease gene was localized, RFLP analysis of other families could reveal who was at risk for the disease, or who was likely to be a carrier of the mutant genes. RFLP test is used in identification and differentiation of organisms by analyzing unique patterns in genome. It is also used in identification of recombination rate in the loci between restriction sites.

RFLP analysis was also the basis for early methods of genetic fingerprinting, useful in the identification of samples retrieved from crime scenes, in the determination of paternity, and in the characterization of genetic diversity or breeding patterns in animal populations.

Alternatives

The technique for RFLP analysis is, however, slow and cumbersome. It requires a large amount of sample DNA, and the combined process of probe labeling, DNA fragmentation, electrophoresis, blotting, hybridization, washing, and autoradiography can take up to a month to complete. A limited version of the RFLP method that used oligonucleotide probes was reported in 1985. The results of the Human Genome Project have largely replaced the need for RFLP mapping, and the identification of many single-nucleotide polymorphisms (SNPs) in that project (as well as the direct identification of many disease genes and mutations) has replaced the need for RFLP disease linkage analysis. The analysis of VNTR alleles continues, but is now usually performed by polymerase chain reaction (PCR) methods. For example, the standard protocols for DNA fingerprinting involve PCR analysis of panels of more than a dozen VNTRs.

RFLP is still used in marker-assisted selection. Terminal restriction fragment length polymorphism (TRFLP or sometimes T-RFLP) is a technique initially developed for characterizing bacterial communities in mixed-species samples. The technique has also been applied to other groups including soil fungi. TRFLP works by PCR amplification of DNA using primer pairs that have been labeled with fluorescent tags. The PCR products are then digested using RFLP enzymes and the resulting patterns visualized using a DNA sequencer. The results are analyzed either by simply counting and comparing bands or peaks in the TRFLP profile, or by matching bands from one or more TRFLP runs to a database of known species. The technique is similar in some aspects to temperature gradient or denaturing gradient gel electrophoresis (TGGE and DGGE).

The sequence changes directly involved with an RFLP can also be analyzed more quickly by PCR. Amplification can be directed across the altered restriction site, and the products digested with the restriction enzyme. This method has been called Cleaved Amplified Polymorphic Sequence (CAPS). Alternatively, the amplified segment can be analyzed by allele-specific oligonucleotide (ASO) probes, a process that can often be done by a simple dot blot.

Polymerase Chain Reaction

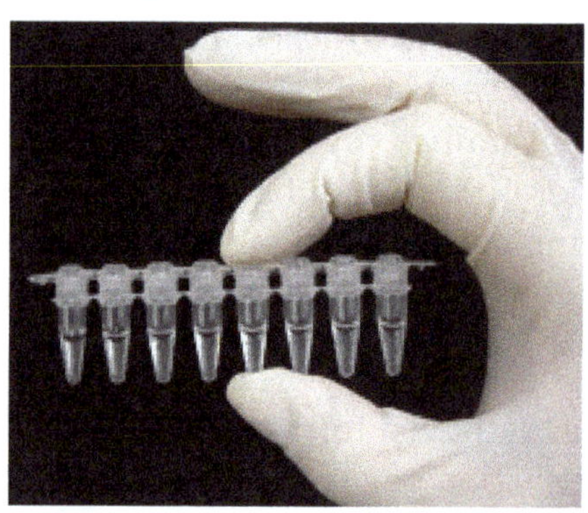

A strip of eight PCR tubes, each containing a 100 μl reaction mixture.

Polymerase chain reaction (PCR) is a method widely used in molecular biology to make several copies of a specific DNA segment. Using PCR, copies of DNA sequences are exponentially amplified to generate thousands to millions of more copies of that particular DNA segment. PCR is now a common and often indispensable technique used in medical laboratory and clinical laboratory research for a broad variety of applications including biomedical research and criminal forensics. PCR was developed by Kary Mullis in 1983 while he was an employee of the Cetus Corporation. He was awarded the Nobel Prize in Chemistry in 1993 (along with Michael Smith) for his work in developing the method.

The vast majority of PCR methods rely on thermal cycling. Thermal cycling exposes reactants to repeated cycles of heating and cooling to permit different temperature-dependent reactions—specifically, DNA melting and enzyme-driven DNA replication. PCR employs two main reagents – primers (which are short single strand DNA fragments known as oligonucleotides that are a complementary sequence to the target DNA region) and a DNA polymerase. In the first step of PCR, the two strands of the DNA double helix are physically separated at a high temperature in a process called DNA melting. In the second step, the temperature is lowered and the primers bind to the complementary sequences of DNA. The two DNA strands then become templates for DNA polymerase to enzymatically assemble a new DNA strand from free nucleotides, the building blocks of DNA. As PCR progresses, the DNA generated is itself used as a template for replication, setting in motion a chain reaction in which the original DNA template is exponentially amplified.

Almost all PCR applications employ a heat-stable DNA polymerase, such as Taq polymerase, an enzyme originally isolated from the thermophilic bacterium *Thermus aquaticus*. If the polymerase used was heat-susceptible, it would denature under the high temperatures of the denaturation step. Before the use of Taq polymerase, DNA polymerase had to be manually added every cycle, which was a tedious and costly process.

Placing a strip of eight PCR tubes into a thermal cycler.

Applications of the technique include DNA cloning for sequencing, gene cloning and manipulation, gene mutagenesis; construction of DNA-based phylogenies, or functional analysis of genes; diagnosis and monitoring of hereditary diseases; amplification of ancient DNA; analysis of genetic fingerprints for DNA profiling (for example, in forensic science and parentage testing); and detection of pathogens in nucleic acid tests for the diagnosis of infectious diseases.

Principles

PCR amplifies a specific region of a DNA strand (the DNA target). Most PCR methods amplify DNA fragments of between 0.1 and 10 kilo base pairs (kbp) in length, although some techniques allow for amplification of fragments up to 40 kbp. The amount of amplified product is determined by the available substrates in the reaction, which become limiting as the reaction progresses.

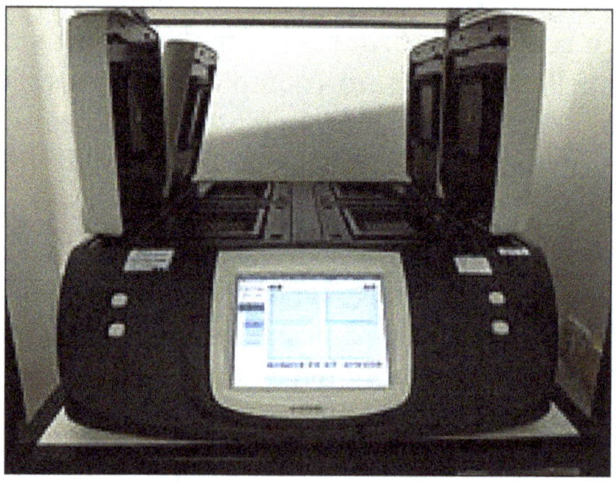

A thermal cycler for PCR.

A basic PCR set-up requires several components and reagents, including a *DNA template* that contains the DNA target region to amplify; a *DNA polymerase*; an enzyme that polymerizes new DNA strands; heat-resistant Taq polymerase is especially common, as it is more likely to remain intact during the high-temperature DNA denaturation process; two DNA *primers* that are complementary to the 3' (three prime) ends of each of the sense and anti-sense strands of the DNA target (DNA polymerase can only bind to and elongate from a double-stranded region of DNA; without primers there is no double-stranded initiation site at which the polymerase can bind); specific primers that are complementary to the DNA target region are selected beforehand, and are often custom-made in a laboratory or purchased from commercial biochemical suppliers; *deoxynucleoside triphosphates*, or dNTPs (sometimes called "deoxynucleotide triphosphates"; nucleotides containing triphosphate groups), the building blocks from which the DNA polymerase synthesizes a new DNA strand; a *buffer solution* providing a suitable chemical environment for optimum activity and stability of the DNA polymerase; *bivalent cations*, typically magnesium (Mg) or manganese (Mn) ions; Mg^{2+} is the most common, but Mn^{2+} can be used for PCR-mediated DNA mutagenesis, as a higher Mn^{2+} concentration increases the error rate during DNA synthesis; and *monovalent cations*, typically potassium (K) ions.

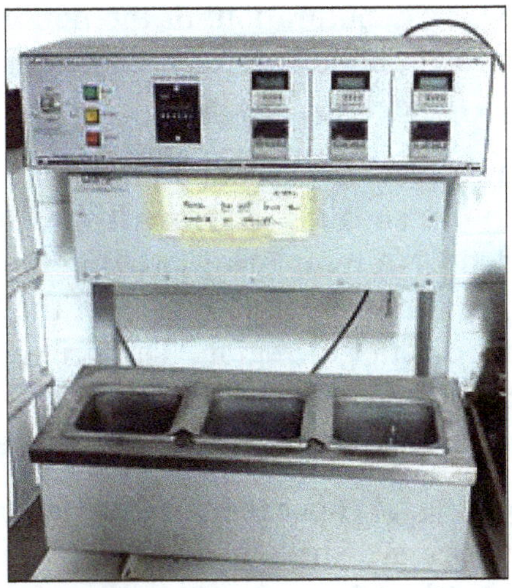

An older, three-temperature thermal cycler for PCR.

The reaction is commonly carried out in a volume of 10–200 μL in small reaction tubes (0.2–0.5 mL volumes) in a thermal cycler. The thermal cycler heats and cools the reaction tubes to achieve the temperatures required at each step of the reaction. Many modern thermal cyclers make use of the Peltier effect, which permits both heating and cooling of the block holding the PCR tubes simply by reversing the electric current. Thin-walled reaction tubes permit favorable thermal conductivity to allow for rapid thermal equilibration. Most thermal cyclers have heated lids to prevent condensation at the top of the reaction tube. Older thermal cyclers lacking a heated lid require a layer of oil on top of the reaction mixture or a ball of wax inside the tube.

Procedure

Typically, PCR consists of a series of 20–40 repeated temperature changes, called thermal cycles, with each cycle commonly consisting of two or three discrete temperature steps. The cycling is often preceded by a single temperature step at a very high temperature (>90 °C (194 °F)), and followed by one hold at the end for final product extension or brief storage. The temperatures used and the length of time they are applied in each cycle depend on a variety of parameters, including the enzyme used for DNA synthesis, the concentration of bivalent ions and dNTPs in the reaction, and the melting temperature (T_m) of the primers. The individual steps common to most PCR methods are as follows:

- Initialization: This step is only required for DNA polymerases that require heat activation by hot-start PCR. It consists of heating the reaction chamber to a temperature of 94–96 °C (201–205 °F), or 98 °C (208 °F) if extremely thermostable polymerases are used, which is then held for 1–10 minutes.
- Denaturation: This step is the first regular cycling event and consists of heating the reaction chamber to 94–98 °C (201–208 °F) for 20–30 seconds. This

causes DNA melting, or denaturation, of the double-stranded DNA template by breaking the hydrogen bonds between complementary bases, yielding two single-stranded DNA molecules.

- Annealing: In the next step, the reaction temperature is lowered to 50–65 °C (122–149 °F) for 20–40 seconds, allowing annealing of the primers to each of the single-stranded DNA templates. Two different primers are typically included in the reaction mixture: one for each of the two single-stranded complements containing the target region. The primers are single-stranded sequences themselves, but are much shorter than the length of the target region, complementing only very short sequences at the 3' end of each strand.

  It is critical to determine a proper temperature for the annealing step because efficiency and specificity are strongly affected by the annealing temperature. This temperature must be low enough to allow for hybridization of the primer to the strand, but high enough for the hybridization to be specific, i.e., the primer should bind *only* to a perfectly complementary part of the strand, and nowhere else. If the temperature is too low, the primer may bind imperfectly. If it is too high, the primer may not bind at all. A typical annealing temperature is about 3–5 °C below the T_m of the primers used. Stable hydrogen bonds between complementary bases are formed only when the primer sequence very closely matches the template sequence. During this step, the polymerase binds to the primer-template hybrid and begins DNA formation.

- Extension/elongation: The temperature at this step depends on the DNA polymerase used; the optimum activity temperature for the thermostable DNA polymerase of Taq (Thermus aquaticus) polymerase is approximately 75–80 °C (167–176 °F), though a temperature of 72 °C (162 °F) is commonly used with this enzyme. In this step, the DNA polymerase synthesizes a new DNA strand complementary to the DNA template strand by adding free dNTPs from the reaction mixture that are complementary to the template in the 5'-to-3' direction, condensing the 5'-phosphate group of the dNTPs with the 3'-hydroxy group at the end of the nascent (elongating) DNA strand. The precise time required for elongation depends both on the DNA polymerase used and on the length of the DNA target region to amplify. As a rule of thumb, at their optimal temperature, most DNA polymerases polymerize a thousand bases per minute. Under optimal conditions (i.e., if there are no limitations due to limiting substrates or reagents), at each extension/elongation step, the number of DNA target sequences is doubled. With each successive cycle, the original template strands plus all newly generated strands become template strands for the next round of elongation, leading to exponential (geometric) amplification of the specific DNA target region.

  The processes of denaturation, annealing and elongation constitute a single cycle. Multiple cycles are required to amplify the DNA target to millions of copies.

The formula used to calculate the number of DNA copies formed after a given number of cycles is 2^n, where n is the number of cycles. Thus, a reaction set for 30 cycles results in 2^{30}, or 1073741824, copies of the original double-stranded DNA target region.

- Final elongation: This single step is optional, but is performed at a temperature of 70–74 °C (158–165 °F) (the temperature range required for optimal activity of most polymerases used in PCR) for 5–15 minutes after the last PCR cycle to ensure that any remaining single-stranded DNA is fully elongated.

- Final hold: The final step cools the reaction chamber to 4–15 °C (39–59 °F) for an indefinite time, and may be employed for short-term storage of the PCR products.

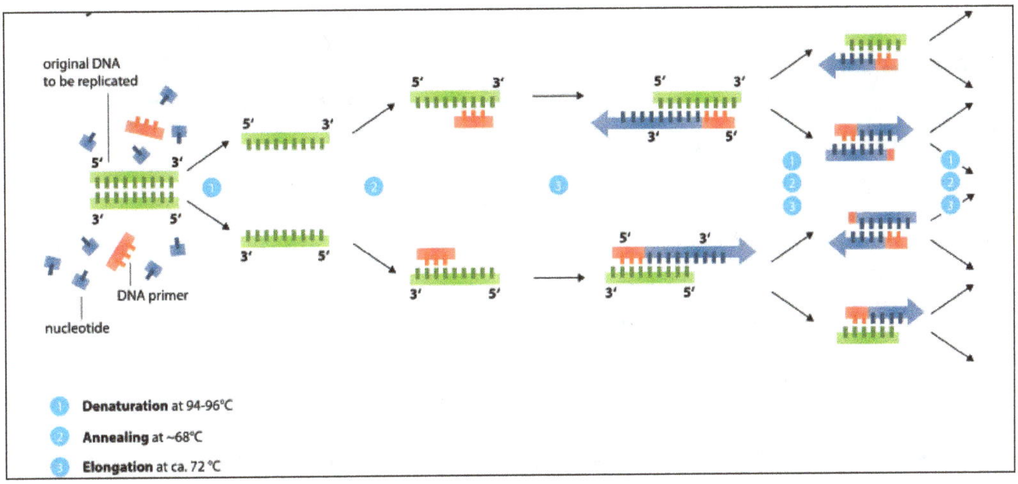

Ethidium bromide-stained PCR products after gel electrophoresis. Two sets of primers were used to amplify a target sequence from three different tissue samples. No amplification is present in sample #1; DNA bands in sample #2 and #3 indicate successful amplification of the target sequence. The gel also shows a positive control, and a DNA ladder containing DNA fragments of defined length for sizing the bands in the experimental PCRs.

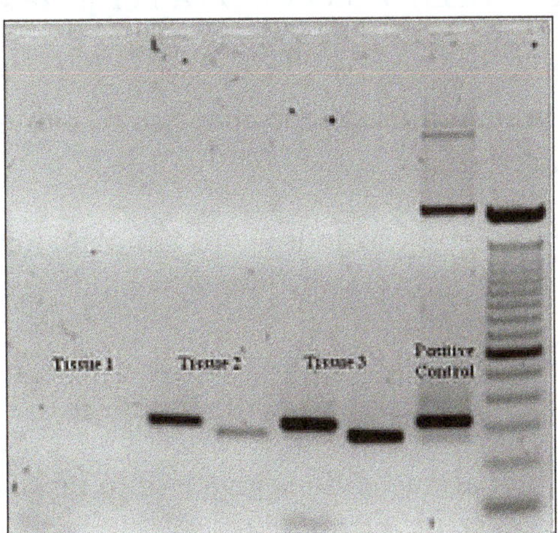

To check whether the PCR successfully generated the anticipated DNA target region (also sometimes referred to as the amplimer or amplicon), agarose gel electrophoresis may be employed for size separation of the PCR products. The size(s) of PCR products is determined by comparison with a DNA ladder, a molecular weight marker which contains DNA fragments of known size run on the gel alongside the PCR products.

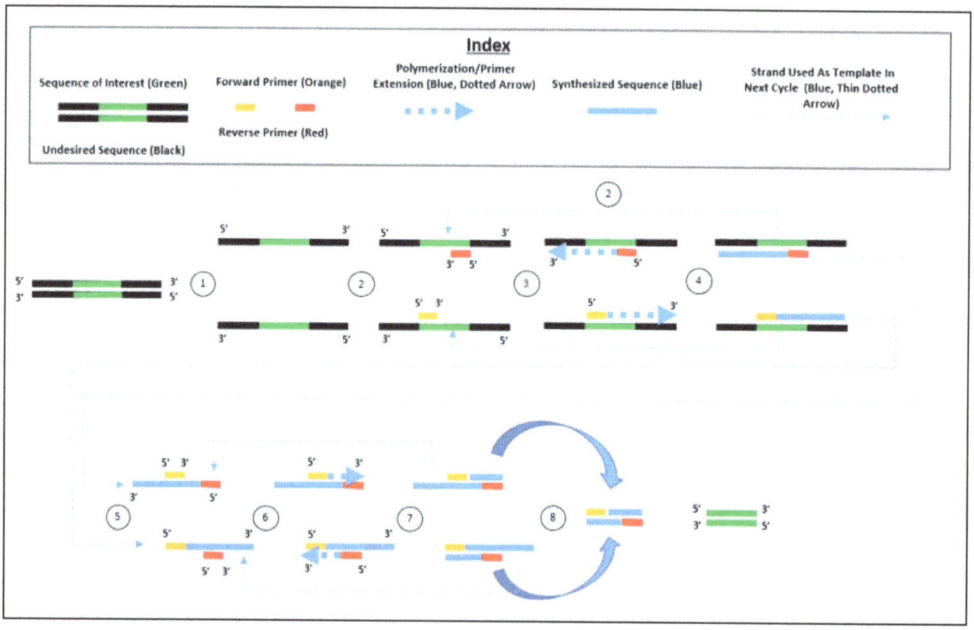

- The DNA double helix is melted apart at T> 90°C and its strands separate.
- The temperature is decreased to slightly below the T. of both the primers being used. Both primers bind to the available strands. These primers are supplied in excess to insure that the strands do not only come back and reanneal to one another.
- Polymerization (extension) occurs via DNA Polymerase in the 5' to 3' direction on each strand.
- Incorporated additional nucleotides give rise to new strands that extend past the sequence of interest.
- The previously polymerized strands act as template for the other primer (if forward primer bound first, reverse primer now binds and vice versa).
- Polymerization occurs via DNA Polymerase in the 5' to 3' direction on each strand, this time ending at the end of the sequence of interest.
- Incorporated additional nucleotides give rise to new strands that only encode the sequence of interest.

- The synthesized strands encoding the sequence of interest anneal to one another to form the end product.

Stages

As with other chemical reactions, the reaction rate and efficiency of PCR are affected by limiting factors. Thus, the entire PCR process can further be divided into three stages based on reaction progress:

- Exponential amplification: At every cycle, the amount of product is doubled (assuming 100% reaction efficiency). After 30 cycles, a single copy of DNA can be increased up to 1,000,000,000 (one billion) copies. In a sense, then, the replication of a discrete strand of DNA is being manipulated in a tube under controlled conditions. The reaction is very sensitive: only minute quantities of DNA must be present.
- Leveling off stage: The reaction slows as the DNA polymerase loses activity and as consumption of reagents, such as dNTPs and primers, causes them to become more limited.
- Plateau: No more product accumulates due to exhaustion of reagents and enzyme.

Optimization

In practice, PCR can fail for various reasons, in part due to its sensitivity to contamination causing amplification of spurious DNA products. Because of this, a number of techniques and procedures have been developed for optimizing PCR conditions. Contamination with extraneous DNA is addressed with lab protocols and procedures that separate pre-PCR mixtures from potential DNA contaminants. This usually involves spatial separation of PCR-setup areas from areas for analysis or purification of PCR products, use of disposable plasticware, and thoroughly cleaning the work surface between reaction setups. Primer-design techniques are important in improving PCR product yield and in avoiding the formation of spurious products, and the usage of alternate buffer components or polymerase enzymes can help with amplification of long or otherwise problematic regions of DNA. Addition of reagents, such as formamide, in buffer systems may increase the specificity and yield of PCR. Computer simulations of theoretical PCR results (Electronic PCR) may be performed to assist in primer design.

Applications

Selective DNA Isolation

PCR allows isolation of DNA fragments from genomic DNA by selective amplification of a specific region of DNA. This use of PCR augments many ways, such as generating hybridization probes for Southern or northern hybridization and DNA cloning, which require larger amounts of DNA, representing a specific DNA region. PCR supplies these techniques with high amounts of pure DNA, enabling analysis of DNA samples even from very small amounts of starting material.

Other applications of PCR include DNA sequencing to determine unknown PCR-amplified sequences in which one of the amplification primers may be used in Sanger sequencing, isolation of a DNA sequence to expedite recombinant DNA technologies involving the insertion of a DNA sequence into a plasmid, phage, or cosmid (depending on size) or the genetic material of another organism. Bacterial colonies *(such as E. coli)* can be rapidly screened by PCR for correct DNA vector constructs. PCR may also be used for genetic fingerprinting; a forensic technique used to identify a person or organism by comparing experimental DNAs through different PCR-based methods.

Some PCR 'fingerprints' methods have high discriminative power and can be used to identify genetic relationships between individuals, such as parent-child or between siblings, and are used in paternity testing. This technique may also be used to determine evolutionary relationships among organisms when certain molecular clocks are used (i.e., the 16S rRNA and recA genes of microorganisms).

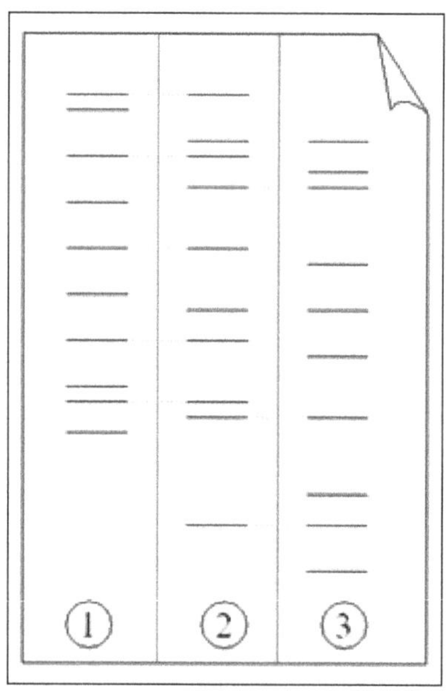

Electrophoresis of PCR-amplified DNA fragments: (1) Father. (2) Child. (3) Mother. The child has inher-ited some, but not all of the fingerprint of each of its parents, giving it a new, unique fingerprint.

Amplification and Quantification of DNA

Because PCR amplifies the regions of DNA that it targets, PCR can be used to analyze extremely small amounts of sample. This is often critical for forensic analysis, when only a trace amount of DNA is available as evidence. PCR may also be used in the analysis of ancient DNA that is tens of thousands of years old. These PCR-based techniques have been successfully used on animals, such as a forty-thousand-year-old mammoth, and also on human DNA, in applications ranging from the analysis of Egyptian mummies to the identification of a Russian tsar and the body of English king Richard III.

Quantitative PCR or Real Time PCR (qPCR, not to be confused with RT-PCR) methods allow the estimation of the amount of a given sequence present in a sample—a technique often applied to quantitatively determine levels of gene expression. Quantitative PCR is an established tool for DNA quantification that measures the accumulation of DNA product after each round of PCR amplification.

qPCR allows the quantification and detection of a specific DNA sequence in real time since it measures concentration while the synthesis process is taking place. There are two methods for simultaneous detection and quantification. The first method consists of using fluorescent dyes that are retained nonspecifically in between the double strands. The second method involves probes that code for specific sequences and are fluorescently labeled. Detection of DNA using these methods can only be seen after the hybridization of probes with its complementary DNA takes place. An interesting technique combination is real-time PCR and reverse transcription. This sophisticated technique, called RT-qPCR, allows for the quantification of a small quantity of RNA. Through this combined technique, mRNA is converted to cDNA, which is further quantified using qPCR. This technique lowers the possibility of error at the end point of PCR, increasing chances for detection of genes associated with genetic diseases such as cancer. Laboratories use RT-qPCR for the purpose of sensitively measuring gene regulation.

Medical and Diagnostic Applications

Prospective parents can be tested for being genetic carriers, or their children might be tested for actually being affected by a disease. DNA samples for prenatal testing can be obtained by amniocentesis, chorionic villus sampling, or even by the analysis of rare fetal cells circulating in the mother's bloodstream. PCR analysis is also essential to preimplantation genetic diagnosis, where individual cells of a developing embryo are tested for mutations.

- PCR can also be used as part of a sensitive test for *tissue typing*, vital to organ transplantation. As of 2008, there is even a proposal to replace the traditional antibody-based tests for blood type with PCR-based tests.
- Many forms of cancer involve alterations to *oncogenes*. By using PCR-based tests to study these mutations, therapy regimens can sometimes be individually customized to a patient. PCR permits early diagnosis of malignant diseases such as leukemia and lymphomas, which is currently the highest-developed in cancer research and is already being used routinely. PCR assays can be performed directly on genomic DNA samples to detect translocation-specific malignant

cells at a sensitivity that is at least 10,000 fold higher than that of other methods. PCR is very useful in the medical field since it allows for the isolation and amplification of tumor suppressors. Quantitative PCR for example, can be used to quantify and analyze single cells, as well as recognize DNA, mRNA and protein confirmations and combinations.

Infectious Disease Applications

PCR allows for rapid and highly specific diagnosis of infectious diseases, including those caused by bacteria or viruses. PCR also permits identification of non-cultivatable or slow-growing microorganisms such as mycobacteria, anaerobic bacteria, or viruses from tissue culture assays and animal models. The basis for PCR diagnostic applications in microbiology is the detection of infectious agents and the discrimination of non-pathogenic from pathogenic strains by virtue of specific genes.

Characterization and detection of infectious disease organisms have been revolutionized by PCR in the following ways:

- The *human immunodeficiency virus* (or *HIV*), is a difficult target to find and eradicate. The earliest tests for infection relied on the presence of antibodies to the virus circulating in the bloodstream. However, antibodies don't appear until many weeks after infection, maternal antibodies mask the infection of a newborn, and therapeutic agents to fight the infection don't affect the antibodies. PCR tests have been developed that can detect as little as one viral genome among the DNA of over 50,000 host cells. Infections can be detected earlier, donated blood can be screened directly for the virus, newborns can be immediately tested for infection, and the effects of antiviral treatments can be quantified.
- Some disease organisms, such as that for *tuberculosis*, are difficult to sample from patients and slow to be grown in the laboratory. PCR-based tests have allowed detection of small numbers of disease organisms (both live or dead), in convenient samples. Detailed genetic analysis can also be used to detect antibiotic resistance, allowing immediate and effective therapy. The effects of therapy can also be immediately evaluated.
- The spread of a *disease organism* through populations of domestic or wild animals can be monitored by PCR testing. In many cases, the appearance of new virulent sub-types can be detected and monitored. The sub-types of an organism that were responsible for earlier epidemics can also be determined by PCR analysis.
- Viral DNA can be detected by PCR: The primers used must be specific to the targeted sequences in the DNA of a virus, and PCR can be used for diagnostic analyses or DNA sequencing of the viral genome. The high sensitivity of PCR permits virus detection soon after infection and even before the onset of disease.

Such early detection may give physicians a significant lead time in treatment. The amount of virus ("viral load") in a patient can also be quantified by PCR-based DNA quantitation techniques.

- Diseases such as pertussis (or whooping cough) are cause by the bacteria Bordetella pertussis. This bacteria is marked by a serious acute respiratory infection that affects various animals and humans and has led to the deaths of many young children. The pertussis toxin is a protein exotoxin that binds to cell receptors by two dimers and reacts with different cell types such as T lymphocytes which plays a role in cell immunity. PCR is an important testing tool that can detect the sequences that are within the pertussis toxin gene. This is because PCR has a high sensitivity for the toxin and has demonstrated a rapid turnaround time. PCR is very efficient for diagnosing pertussis when compared to culture.

Forensic Applications

The development of PCR-based genetic (or DNA) fingerprinting protocols has seen widespread application in forensics:

- In its most discriminating form, *genetic fingerprinting* can uniquely discriminate any one person from the entire population of the world. Minute samples of DNA can be isolated from a crime scene, and compared to that from suspects, or from a DNA database of earlier evidence or convicts. Simpler versions of these tests are often used to rapidly rule out suspects during a criminal investigation. Evidence from decades-old crimes can be tested, confirming or exonerating the people originally convicted.

- Forensic DNA typing has been an effective way of identifying or exonerating criminal suspects due to analysis of evidence discovered at a crime scene. The human genome has many repetitive regions that can be found within gene sequences or in non-coding regions of the genome. Specifically, up to 40% of human DNA is repetitive. There are two distinct categories for these repetitive, non-coding regions in the genome. The first category is called variable number tandem repeats (VNTR), which are 10–100 base pairs long and the second category is called short tandem repeats (STR) and these consist of repeated 2–10 base pair sections. PCR is used to amplify several well-known VNTRs and STRs using primers that flank each of the repetitive regions. The sizes of the fragments obtained from any individual for each of the STRs will indicate which alleles are present. By analyzing several STRs for an individual, a set of alleles for each person will be found that statistically is likely to be unique. Researchers have identified the complete sequence of the human genome. This sequence can be easily accessed through the NCBI website and is used in many real-life applications. For example, the FBI has compiled a set of DNA marker sites used for identification, and these are called the Combined DNA Index System (CODIS) DNA database. Using this database enables statistical analysis to be used to determine the probability that a DNA sample will match. PCR is a very powerful and significant analytical tool to use for forensic DNA typing because researchers only need a very small amount of the target DNA to be used for analysis. For example, a single human hair with attached hair follicle has enough DNA to conduct the analysis. Similarly, a few sperm, skin samples from under the fingernails, or a small amount of blood can provide enough DNA for conclusive analysis.

- Less discriminating forms of DNA fingerprinting can help in *DNA paternity testing*, where an individual is matched with their close relatives. DNA from unidentified human remains can be tested, and compared with that from possible parents, siblings, or children. Similar testing can be used to confirm the biological parents of an adopted (or kidnapped) child. The actual biological father of a newborn can also be confirmed (or ruled out).
- The PCR AMGX/AMGY design has been shown to not only facilitating in amplifying DNA sequences from a very minuscule amount of genome. However it can also be used for real time sex determination from forensic bone samples. This provides us with a powerful and effective way to determine the sex of not only ancient specimens but also current suspects in crimes.

Research Applications

PCR has been applied to many areas of research in molecular genetics:

- PCR allows rapid production of short pieces of DNA, even when not more than the sequence of the two primers is known. This ability of PCR augments many methods, such as generating *hybridization probes* for Southern or northern blot hybridization. PCR supplies these techniques with large amounts of pure DNA, sometimes as a single strand, enabling analysis even from very small amounts of starting material.
- The task of *DNA sequencing* can also be assisted by PCR. Known segments of DNA can easily be produced from a patient with a genetic disease mutation. Modifications to the amplification technique can extract segments from a completely unknown genome, or can generate just a single strand of an area of interest.
- PCR has numerous applications to the more traditional process of *DNA cloning*. It can extract segments for insertion into a vector from a larger genome, which may be only available in small quantities. Using a single set of 'vector primers', it can also analyze or extract fragments that have already been inserted into vectors. Some alterations to the PCR protocol can *generate mutations* (general or site-directed) of an inserted fragment.

- *Sequence-tagged sites* is a process where PCR is used as an indicator that a particular segment of a genome is present in a particular clone. The Human Genome Project found this application vital to mapping the cosmid clones they were sequencing, and to coordinating the results from different laboratories.
- An exciting application of PCR is the phylogenic analysis of DNA from *ancient sources*, such as that found in the recovered bones of Neanderthals, from frozen tissues of mammoths, or from the brain of Egyptian mummies. Have been amplified and sequenced. In some cases the highly degraded DNA from these sources might be reassembled during the early stages of amplification.
- A common application of PCR is the study of patterns of *gene expression*. Tissues (or even individual cells) can be analyzed at different stages to see which genes have become active, or which have been switched off. This application can also use quantitative PCR to quantitate the actual levels of expression.
- The ability of PCR to simultaneously amplify several loci from individual sperm has greatly enhanced the more traditional task of *genetic mapping* by studying chromosomal crossovers after meiosis. Rare crossover events between very close loci have been directly observed by analyzing thousands of individual sperms. Similarly, unusual deletions, insertions, translocations, or inversions can be analyzed, all without having to wait (or pay) for the long and laborious processes of fertilization, embryogenesis, etc.
- Site-directed mutagenesis: PCR can be used to create mutant genes with mutations chosen by scientists at will. These mutations can be chosen in order to understand how proteins accomplish their functions, and to change or improve protein function.

Advantages

PCR has a number of advantages. It is fairly simple to understand and to use, and produces results rapidly. The technique is highly sensitive with the potential to produce millions to billions of copies of a specific product for sequencing, cloning, and analysis. qRT-PCR shares the same advantages as the PCR, with an added advantage of quantification of the synthesized product. Therefore, it has its uses to analyze alterations of gene expression levels in tumors, microbes, or other disease states.

PCR is a very powerful and practical research tool. The sequencing of unknown etiologies of many diseases are being figured out by the PCR. The technique can help identify the sequence of previously unknown viruses related to those already known and thus give us a better understanding of the disease itself. If the procedure can be further simplified and sensitive non radiometric detection systems can be developed, the PCR will assume a prominent place in the clinical laboratory for years to come.

Limitations

One major limitation of PCR is that prior information about the target sequence is necessary in order to generate the primers that will allow its selective amplification. This means that, typically, PCR users must know the precise sequence(s) upstream of the target region on each of the two single-stranded templates in order to ensure that the DNA polymerase properly binds to the primer-template hybrids and subsequently generates the entire target region during DNA synthesis.

Like all enzymes, DNA polymerases are also prone to error, which in turn causes mutations in the PCR fragments that are generated.

Another limitation of PCR is that even the smallest amount of contaminating DNA can be amplified, resulting in misleading or ambiguous results. To minimize the chance of contamination, investigators should reserve separate rooms for reagent preparation, the PCR, and analysis of product. Reagents should be dispensed into single-use aliquots. Pipetters with disposable plungers and extra-long pipette tips should be routinely used.

Short Tandem Repeats

A short tandem repeat (STR) in DNA occurs when a pattern of two or more nucleotides are repeated and the repeated sequences are directly adjacent to each other. An STR is also known as a microsatellite. The pattern can range in length from 2 to 16 base pairs (bp) and is typically in the non-coding intron region. A short tandem repeat polymorphism (STRP) occurs when homologous STR loci differ in the number of repeats between individuals. By identifying repeats of a specific sequence at specific locations in the genome, it is possible to create a genetic profile of an individual. Autosomal STR analysis has become the prevalent analysis method for determining genetic profiles in forensic cases. Y-chromosome STRs (Y-STRs) on the Y chromosome) are used in genealogical DNA testing in surname DNA projects.

Amplified Fragment Length Polymorphism

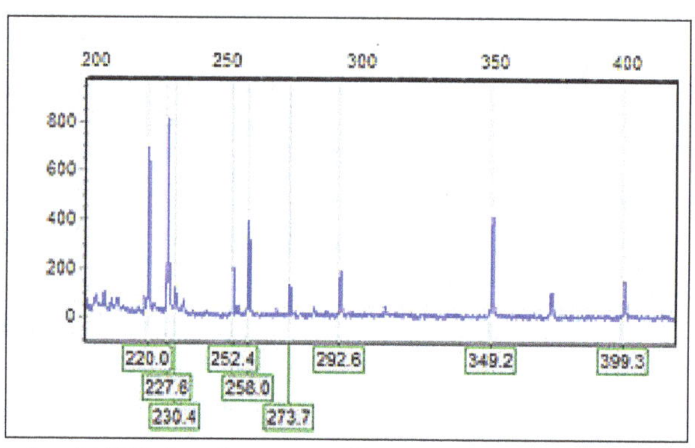

Example of AFLP data from a capillary electrophoresis instrument.

AFLP-PCR or just AFLP is a PCR-based tool used in genetics research, DNA fingerprinting, and in the practice of genetic engineering. Developed in the early 1990s by Keygene, AFLP uses restriction enzymes to digest genomic DNA, followed by ligation of adaptors to the sticky ends of the restriction fragments. A subset of the restriction fragments is then selected to be amplified. This selection is achieved by using primers complementary to the adaptor sequence, the restriction site sequence and a few nucleotides inside the restriction site fragments. The amplified fragments are separated and visualized on denaturing on agarose gel electrophoresis , either through autoradiography or fluorescence methodologies, or via automated capillary sequencing instruments.

Although AFLP is commonly referred to as "Amplified fragment length polymorphism", the resulting data are not scored as length polymorphisms, but instead as presence-absence polymorphisms.

AFLP-PCR is a highly sensitive method for detecting polymorphisms in DNA. The technique was originally described by Vos and Zabeau in 1993. In detail, the procedure of this technique is divided into three steps:

- Digestion of total cellular DNA with one or more restriction enzymes and ligation of restriction half-site specific adaptors to all restriction fragments.
- Selective amplification of some of these fragments with two PCR primers that have corresponding adaptor and restriction site specific sequences.
- Electrophoretic separation of amplicons on a gel matrix, followed by visualisation of the band pattern.

Applications

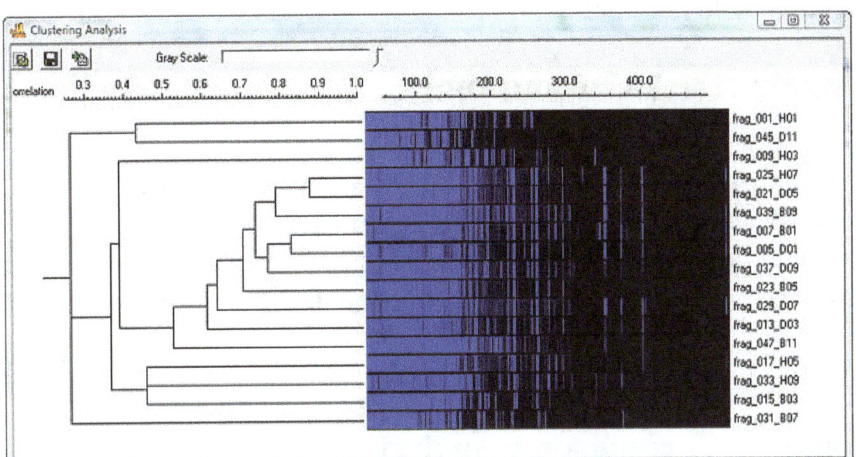

AFLP phylogeny analysis using a dendrogram.

The AFLP technology has the capability to detect various polymorphisms in different genomic regions simultaneously. It is also highly sensitive and reproducible. As a result,

AFLP has become widely used for the identification of genetic variation in strains or closely related species of plants, fungi, animals, and bacteria. The AFLP technology has been used in criminal and paternity tests, also to determine slight differences within populations, and in linkage studies to generate maps for quantitative trait locus (QTL) analysis.

There are many advantages to AFLP when compared to other marker technologies including randomly amplified polymorphic DNA (RAPD), restriction fragment length polymorphism (RFLP), and microsatellites. AFLP not only has higher reproducibility, resolution, and sensitivity at the whole genome level compared to other techniques, but it also has the capability to amplify between 50 and 100 fragments at one time. In addition, no prior sequence information is needed for amplification. As a result, AFLP has become extremely beneficial in the study of taxa including bacteria, fungi, and plants, where much is still unknown about the genomic makeup of various organisms.

Y Chromosome Analysis

Y chromosome analysis is a useful technique for analysing DNA that can be likened in one sense to studying male surnames. Think about the way that male surnames are passed down from one generation to another and continue on through sons. This mechanism is a simplistic representation of Y chromosomes. A son inherits a Y chromosome from his biological father and he also inherits an X chromosome from his biological mother. Conversely, a female would inherit an X chromosome from her biological mother and an X chromosome from her biological father. As such, when scientists study Y chromosomes, they are studying these chromosomes as they are inherited over time through males in a familial line. This type of DNA analysis has important ramifications for scientists wishing to investigate the familial ties between male members.

Understanding the Y Chromosome

Over time, Y chromosomes will slowly change. As a Y chromosome is replicated from one generation to another, the process is one whereby mistakes do not typically occur. If, however, mistakes were to occur - particularly in a large number - then the consequences would be dire. As such, while most of the time the process is an error-free one, there are errors that can occasionally occur and these errors will be passed from a father to his son. Thus, one critical application of Y chromosome analysis is to study genetic mutations.

Conditions of Y Chromosome Analysis

While we can learn a great deal from the analysis of Y chromosomes, this knowledge is dependent on several factors. One factor is the decision of how many loci will be compared between the Y chromosomes. There are a minimum number of markers needed to collect the Y chromosomes into biological family sets. As a larger number of markers are utilised, the result is that lineages can be investigated more thoroughly. Our knowledge of Y chromosomes is also dependent on picking the correct loci to analyse for mutations on the Y chromosome. Loci have varying mutation rates, which means that picking the correct one is vital. It is also important to note that the greater the number of people analysed in a Y chromosome study, the better the result.

Using and Applying Y Chromosome Analysis

Y chromosome analysis has many useful applications in several different fields. It can be used to provide evidence of any familial relationship between males, which is particularly valuable when there is a lack of paper evidence to establish a link. One important application for Y chromosome analysis is to help families who have immigrated frequently to many countries and the paperwork is challenging to find or perhaps does not even exist anymore. Y chromosome analysis is also useful for application to families in one area that are not even aware they are related because of surname differences, for example. Therefore, this type of analysis can establish familial links.

While there are many different techniques for analysing DNA, Y chromosome analysis clearly has important and useful applications for a number of fields. It is less suited to criminal investigations but provides many benefits for investigating familial relationships, which is particularly valuable for families wishing to identify other members.

DNA Sequencing

DNA sequencing is the process of determining the nucleic acid sequence – the order of nucleotides in DNA. It includes any method or technology that is used to determine the order of the four bases: adenine, guanine, cytosine, and thymine. The advent of rapid DNA sequencing methods has greatly accelerated biological and medical research and discovery.

Knowledge of DNA sequences has become indispensable for basic biological research, and in numerous applied fields such as medical diagnosis, biotechnology, forensic biology, virology and biological systematics. Comparing healthy and mutated DNA sequences can diagnose different diseases including various cancers, characterize antibody repertoire, and can be used to guide patient treatment. Having a quick way to sequence DNA allows for faster and more individualized medical care to be administered, and for more organisms to be identified and cataloged.

The rapid speed of sequencing attained with modern DNA sequencing technology has been instrumental in the sequencing of complete DNA sequences, or genomes, of numerous types and species of life, including the human genome and other complete DNA sequences of many animal, plant, and microbial species.

The first DNA sequences were obtained in the early 1970s by academic researchers using laborious methods based on two-dimensional chromatography. Following the development of fluorescence-based sequencing methods with a DNA sequencer, DNA sequencing has become easier and orders of magnitude faster.

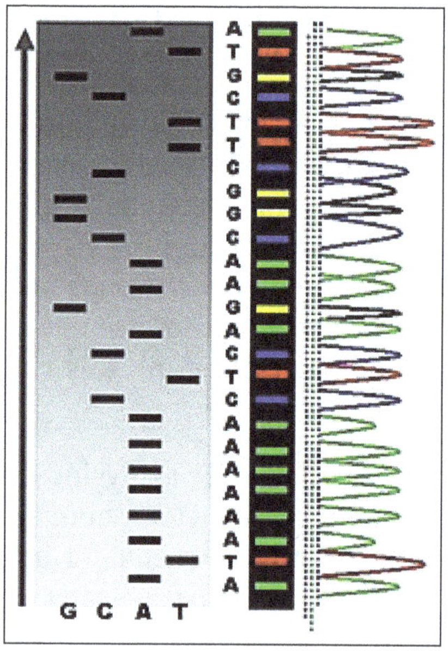

An example of the results of automated chain-termination DNA sequencing.

Applications

DNA sequencing may be used to determine the sequence of individual genes, larger genetic regions (i.e. clusters of genes or operons), full chromosomes, or entire genomes of any organism. DNA sequencing is also the most efficient way to indirectly sequence RNA or proteins (via their open reading frames). In fact, DNA sequencing has become a key technology in many areas of biology and other sciences such as medicine, forensics, and anthropology.

Molecular Biology

Sequencing is used in molecular biology to study genomes and the proteins they encode. Information obtained using sequencing allows researchers to identify changes in genes, associations with diseases and phenotypes, and identify potential drug targets.

Evolutionary Biology

Since DNA is an informative macromolecule in terms of transmission from one generation to another, DNA sequencing is used in evolutionary biology to study how different organisms are related and how they evolved.

Metagenomics

The field of metagenomics involves identification of organisms present in a body of water, sewage, dirt, debris filtered from the air, or swab samples from organisms. Knowing which organisms are present in a particular environment is critical to research in ecology, epidemiology, microbiology, and other fields. Sequencing enables researchers to determine which types of microbes may be present in a microbiome.

Medicine

Medical technicians may sequence genes (or, theoretically, full genomes) from patients to determine if there is risk of genetic diseases. This is a form of genetic testing, though some genetic tests may not involve DNA sequencing. Also, DNA sequencing may be useful for determining a specific bacteria, to allow for more precise antibiotics treatments, hereby reducing the risk of creating antimicrobial resistance in bacteria populations.

Forensics

DNA sequencing may be used along with DNA profiling methods for forensic identification and paternity testing. DNA testing has evolved tremendously in the last few decades to ultimately link a DNA print to what is under investigation. The DNA patterns in fingerprint, saliva, hair follicles, etc. uniquely separate each living organism from another. Testing DNA is a technique which can detect specific genomes in a DNA strand to produce a unique and individualized pattern.

The Four Canonical Bases

The canonical structure of DNA has four bases: thymine (T), adenine (A), cytosine (C), and guanine (G). DNA sequencing is the determination of the physical order of these bases in a molecule of DNA. However, there are many other bases that may be present in a molecule. In some viruses (specifically, bacteriophage), cytosine may be replaced by hydroxy methyl or hydroxy methyl glucose cytosine. In mammalian DNA, variant bases with methyl groups or phosphosulfate may be found. Depending on the sequencing technique, a particular modification, e.g., the 5mC (5 methyl cytosine) common in humans, may or may not be detected.

Basic Methods

Maxam-Gilbert Sequencing

Allan Maxam and Walter Gilbert published a DNA sequencing method in 1977 based on chemical modification of DNA and subsequent cleavage at specific bases. Also known as chemical sequencing, this method allowed purified samples of double-stranded DNA to be used without further cloning. This method's use of radioactive labeling and its technical complexity discouraged extensive use after refinements in the Sanger methods had been made.

Maxam-Gilbert sequencing requires radioactive labeling at one 5' end of the DNA and purification of the DNA fragment to be sequenced. Chemical treatment then generates breaks at a small proportion of one or two of the four nucleotide bases in each of four reactions (G, A+G, C, C+T). The concentration of the modifying chemicals is controlled to introduce on average one modification per DNA molecule. Thus a series of labeled fragments is generated, from the radiolabeled end to the first "cut" site in each molecule. The fragments in the four reactions are electrophoresed side by side in denaturing acrylamide gels for size separation. To visualize the fragments, the gel is exposed to X-ray film for autoradiography, yielding a series of dark bands each corresponding to a radiolabeled DNA fragment, from which the sequence may be inferred.

Chain-termination Methods

The chain-termination method developed by Frederick Sanger and coworkers in 1977 soon became the method of choice, owing to its relative ease and reliability. When invented, the chain-terminator method used fewer toxic chemicals and lower amounts of radioactivity than the Maxam and Gilbert method. Because of its comparative ease, the Sanger method was soon automated and was the method used in the first generation of DNA sequencers.

Sanger sequencing is the method which prevailed from the 1980s until the mid-2000s. Over that period, great advances were made in the technique, such as fluorescent labelling, capillary electrophoresis, and general automation. These developments allowed much more efficient sequencing, leading to lower costs. The Sanger method, in mass production form, is the technology which produced the first human genome in 2001, ushering in the age of genomics. However, later in the decade, radically different approaches reached the market, bringing the cost per genome down from $100 million in 2001 to $10,000 in 2011.

Advanced Methods and de Novo Sequencing

Large-scale sequencing often aims at sequencing very long DNA pieces, such as whole chromosomes, although large-scale sequencing can also be used to generate very large numbers of short sequences, such as found in phage display. For longer targets such as chromosomes, common approaches consist of cutting (with restriction enzymes) or shearing (with mechanical forces) large DNA fragments into shorter DNA fragments. The fragmented DNA may then be cloned into a DNA vector and amplified in a bacterial host such as *Escherichia coli*. Short DNA fragments purified from individual bacterial colonies are individually sequenced and assembled electronically into one long, contiguous sequence. Studies have shown that adding a size selection step to collect DNA fragments of uniform size can improve sequencing efficiency and accuracy of the genome assembly. In these studies, automated sizing has proven to be more reproducible and precise than manual gel sizing.

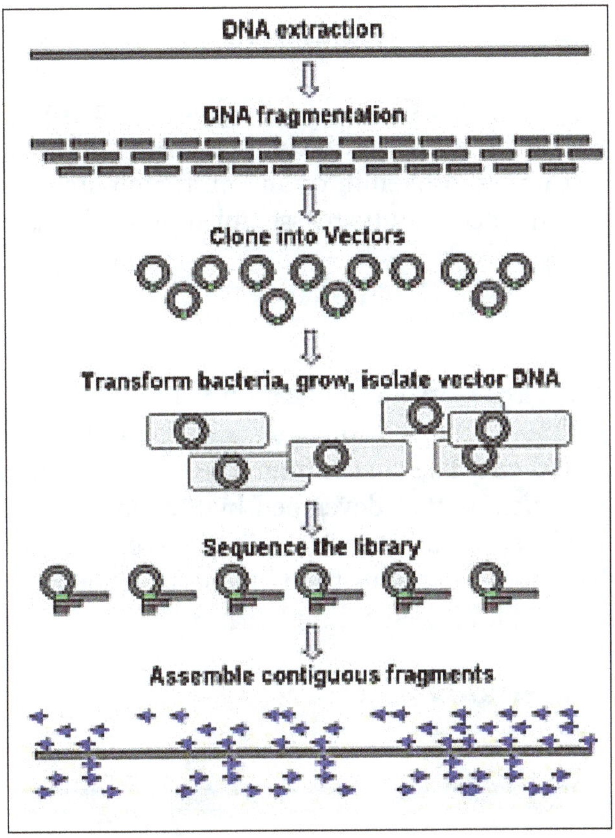

Genomic DNA is fragmented into random pieces and cloned as a bacterial library. DNA from individual bacterial clones is sequenced and the sequence is assembled by using overlapping DNA regions.

The term "*de novo* sequencing" specifically refers to methods used to determine the sequence of DNA with no previously known sequence. *De novo* translates from Latin as "from the beginning". Gaps in the assembled sequence may be filled by primer walking. The different strategies have different tradeoffs in speed and accuracy; shotgun methods are often used for sequencing large genomes, but its assembly is complex and difficult, particularly with sequence repeats often causing gaps in genome assembly.

Most sequencing approaches use an *in vitro* cloning step to amplify individual DNA molecules, because their molecular detection methods are not sensitive enough for single molecule sequencing. Emulsion PCR isolates individual DNA molecules along with primer-coated beads in aqueous droplets within an oil phase. A polymerase chain reaction (PCR) then coats each bead with clonal copies of the DNA molecule followed by immobilization for later sequencing. Emulsion PCR is used in the methods developed by Marguilis et al. (commercialized by 454 Life Sciences), Shendure and Porreca et al.

(also known as "polony sequencing") and SOLiD sequencing, (developed by Agencourt, later Applied Biosystems, now Life Technologies). Emulsion PCR is also used in the GemCode and Chromium platforms developed by 10x Genomics.

Shotgun Sequencing

Shotgun sequencing is a sequencing method designed for analysis of DNA sequences longer than 1000 base pairs, up to and including entire chromosomes. This method requires the target DNA to be broken into random fragments. After sequencing individual fragments, the sequences can be reassembled on the basis of their overlapping regions.

Bridge PCR

Another method for *in vitro* clonal amplification is bridge PCR, in which fragments are amplified upon primers attached to a solid surface and form "DNA colonies" or "DNA clusters". This method is used in the Illumina Genome Analyzer sequencers. Single-molecule methods, such as that developed by Stephen Quake's laboratory (later commercialized by Helicos) are an exception: they use bright fluorophores and laser excitation to detect base addition events from individual DNA molecules fixed to a surface, eliminating the need for molecular amplification.

High-throughput Methods

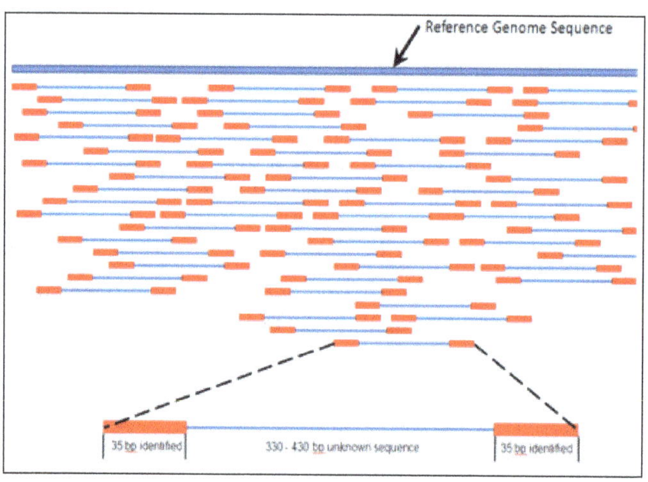

Multiple, fragmented sequence reads must be assembled together on the basis of their overlapping areas.

High-throughput, or next-generation, sequencing applies to genome sequencing, genome resequencing, transcriptome profiling (RNA-Seq), DNA-protein interactions (ChIP-sequencing), and epigenome characterization. Resequencing is necessary, because the genome of a single individual of a species will not indicate all of the genome variations among other individuals of the same species.

The high demand for low-cost sequencing has driven the development of high-throughput sequencing technologies that parallelize the sequencing process, producing thousands or millions of sequences concurrently. High-throughput sequencing technologies are intended to lower the cost of DNA sequencing beyond what is possible with standard dye-terminator methods. In ultra-high-throughput sequencing as many as 500,000 sequencing-by-synthesis operations may be run in parallel. Such technologies led to the ability to sequence an entire human genome in as little as one day. As of 2019, corporate leaders in the development of high-throughput sequencing products included Illumina, Qiagen and ThermoFisher Scientific.

Massively Parallel Signature Sequencing (MPSS)

The first of the high-throughput sequencing technologies, massively parallel signature sequencing (or MPSS), was developed in the 1990s at Lynx Therapeutics, a company founded in 1992 by Sydney Brenner and Sam Eletr. MPSS was a bead-based method that used a complex approach of adapter ligation followed by adapter decoding, reading the sequence in increments of four nucleotides. This method made it susceptible to sequence-specific bias or loss of specific sequences. Because the technology was so complex, MPSS was only performed 'in-house' by Lynx Therapeutics and no DNA sequencing machines were sold to independent laboratories. Lynx Therapeutics merged with Solexa (later acquired by Illumina) in 2004, leading to the development of sequencing-by-synthesis, a simpler approach acquired from Manteia Predictive Medicine, which rendered MPSS obsolete. However, the essential properties of the MPSS output were typical of later high-throughput data types, including hundreds of thousands of short DNA sequences. In the case of MPSS, these were typically used for sequencing cDNA for measurements of gene expression levels.

Polony Sequencing

The polony sequencing method, developed in the laboratory of George M. Church at Harvard, was among the first high-throughput sequencing systems and was used to sequence a full *E. coli* genome in 2005. It combined an in vitro paired-tag library with emulsion PCR, an automated microscope, and ligation-based sequencing chemistry to sequence an *E. coli* genome at an accuracy of >99.9999% and a cost approximately 1/9 that of Sanger sequencing. The technology was licensed to Agencourt Biosciences, subsequently spun out into Agencourt Personal Genomics, and eventually incorporated into the Applied Biosystems SOLiD platform. Applied Biosystems was later acquired by Life Technologies, now part of Thermo Fisher Scientific.

454 Pyrosequencing

A parallelized version of pyrosequencing was developed by 454 Life Sciences, which has since been acquired by Roche Diagnostics. The method amplifies DNA inside water droplets in an oil solution (emulsion PCR), with each droplet containing a single

DNA template attached to a single primer-coated bead that then forms a clonal colony. The sequencing machine contains many picoliter-volume wells each containing a single bead and sequencing enzymes. Pyrosequencing uses luciferase to generate light for detection of the individual nucleotides added to the nascent DNA, and the combined data are used to generate sequence reads. This technology provides intermediate read length and price per base compared to Sanger sequencing on one end and Solexa and SOLiD on the other.

Illumina (Solexa) Sequencing

Solexa, now part of Illumina, was founded by Shankar Balasubramanian and David Klenerman in 1998, and developed a sequencing method based on reversible dye-terminators technology, and engineered polymerases. The reversible terminated chemistry concept was invented by Bruno Canard and Simon Sarfati at the Pasteur Institute in Paris. It was developed internally at Solexa by those named on the relevant patents. In 2004, Solexa acquired the company Manteia Predictive Medicine in order to gain a massively parallel sequencing technology invented in 1997 by Pascal Mayer and Laurent Farinelli. It is based on "DNA clusters" or "DNA colonies", which involves the clonal amplification of DNA on a surface. The cluster technology was co-acquired with Lynx Therapeutics of California. Solexa Ltd. later merged with Lynx to form Solexa Inc.

An Illumina HiSeq 2500 sequencer.

Illumina NovaSeq 6000 flow cell.

In this method, DNA molecules and primers are first attached on a slide or flow cell and amplified with polymerase so that local clonal DNA colonies, later coined "DNA clusters", are formed. To determine the sequence, four types of reversible terminator bases

(RT-bases) are added and non-incorporated nucleotides are washed away. A camera takes images of the fluorescently labeled nucleotides. Then the dye, along with the terminal 3' blocker, is chemically removed from the DNA, allowing for the next cycle to begin. Unlike pyrosequencing, the DNA chains are extended one nucleotide at a time and image acquisition can be performed at a delayed moment, allowing for very large arrays of DNA colonies to be captured by sequential images taken from a single camera.

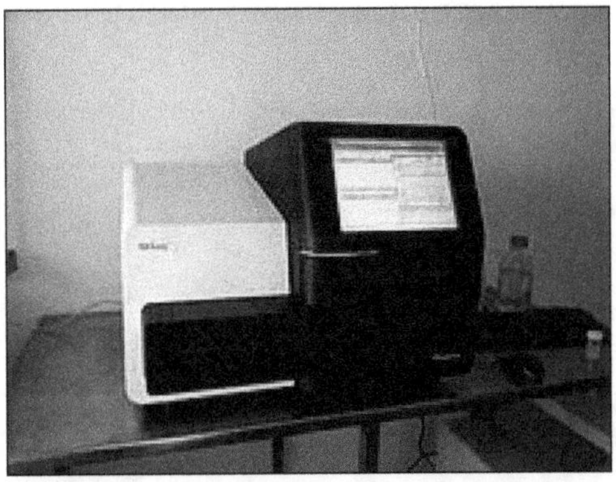

An Illumina MiSeq sequencer.

Decoupling the enzymatic reaction and the image capture allows for optimal throughput and theoretically unlimited sequencing capacity. With an optimal configuration, the ultimately reachable instrument throughput is thus dictated solely by the analog-to-digital conversion rate of the camera, multiplied by the number of cameras and divided by the number of pixels per DNA colony required for visualizing them optimally (approximately 10 pixels/colony). In 2012, with cameras operating at more than 10 MHz A/D conversion rates and available optics, fluidics and enzymatics, throughput can be multiples of 1 million nucleotides/second, corresponding roughly to 1 human genome equivalent at 1x coverage per hour per instrument, and 1 human genome re-sequenced (at approx. 30x) per day per instrument (equipped with a single camera).

Combinatorial Probe Anchor Synthesis (cPAS)

This method is an upgraded modification to combinatorial probe anchor ligation technology (cPAL) described by Complete Genomics which has since become part of Chinese genomics company BGI in 2013. The two companies have refined the technology to allow for longer read lengths, reaction time reductions and faster time to results. In addition, data are now generated as contiguous full-length reads in the standard FASTQ file format and can be used as-is in most short-read-based bioinformatics analysis pipelines.

The two technologies that form the basis for this high-throughput sequencing technology are DNA nanoballs (DNB) and patterned arrays for nanoball attachment to a solid surface. DNA nanoballs are simply formed by denaturing double stranded, adapter ligated libraries and ligating the forward strand only to a splint oligonucleotide to form a ssDNA circle. Faithful copies of the circles containing the DNA insert are produced utilizing Rolling Circle Amplification that generates approximately 300–500 copies. The long strand of ssDNA folds upon itself to produce a three-dimensional nanoball structure that is approximately 220 nm in diameter. Making DNBs replaces the need to generate PCR copies of the library on the flow cell and as such can remove large proportions of duplicate reads, adapter-adapter ligations and PCR induced errors.

A BGI MGISEQ-2000RS sequencer.

The patterned array of positively charged spots is fabricated through photolithography and etching techniques followed by chemical modification to generate a sequencing flow cell. Each spot on the flow cell is approximately 250 nm in diameter, are separated by 700 nm (centre to centre) and allows easy attachment of a single negatively charged DNB to the flow cell and thus reducing under or over-clustering on the flow cell.

Sequencing is then performed by addition of an oligonucleotide probe that attaches in combination to specific sites within the DNB. The probe acts as an anchor that then allows one of four single reversibly inactivated, labelled nucleotides to bind after flowing across the flow cell. Unbound nucleotides are washed away before laser excitation of the attached labels then emit fluorescence and signal is captured by cameras that is converted to a digital output for base calling. The attached base has its terminator and label chemically cleaved at completion of the cycle. The cycle is repeated with another flow of free, labelled nucleotides across the flow cell to allow the next nucleotide to bind and have its signal captured. This process is completed a number of times (usually 50 to 300 times) to determine the sequence of the inserted piece of DNA at a rate of approximately 40 million nucleotides per second as of 2018.

SOLiD Sequencing

Applied Biosystems' (now a Life Technologies brand) SOLiD technology employs sequencing by ligation. Here, a pool of all possible oligonucleotides of a fixed length are labeled according to the sequenced position. Oligonucleotides are annealed and ligated; the preferential ligation by DNA ligase for matching sequences results in a signal informative of the nucleotide at that position. Before sequencing, the DNA is amplified by emulsion PCR. The resulting beads, each containing single copies of the same DNA molecule, are deposited on a glass slide. The result is sequences of quantities and lengths comparable to Illumina sequencing. This sequencing by ligation method has been reported to have some issue sequencing palindromic sequences.

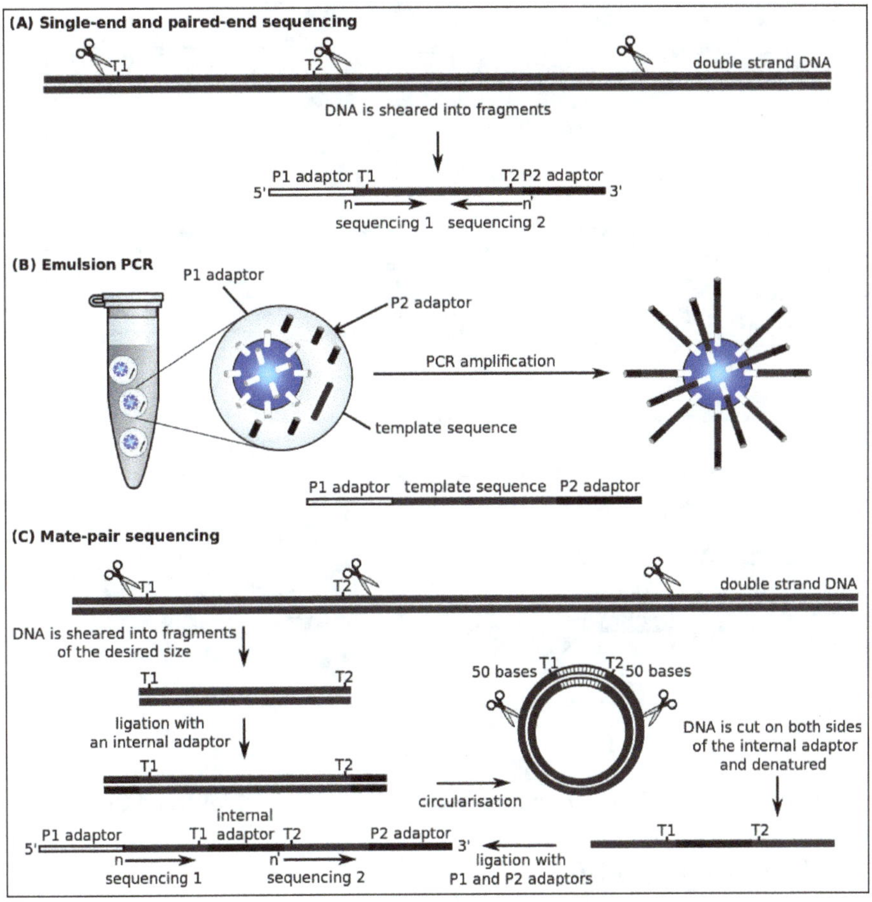

Library preparation for the SOLiD platform.

Ion Torrent Semiconductor Sequencing

Ion Torrent Systems Inc. (now owned by Life Technologies) developed a system based on using standard sequencing chemistry, but with a novel, semiconductor-based detection system. This method of sequencing is based on the detection of hydrogen ions that are released during the polymerisation of DNA, as opposed to the optical methods used in other sequencing systems. A microwell containing a template DNA strand to be sequenced is flooded with a single type of nucleotide. If the introduced nucleotide is complementary to the leading template nucleotide it is incorporated into the growing complementary strand. This causes the release of a hydrogen ion that triggers a hypersensitive ion sensor, which indicates that a reaction has occurred. If homopolymer repeats are present in the template sequence, multiple nucleotides will be incorporated in a single cycle. This leads to a corresponding number of released hydrogens and a proportionally higher electronic signal.

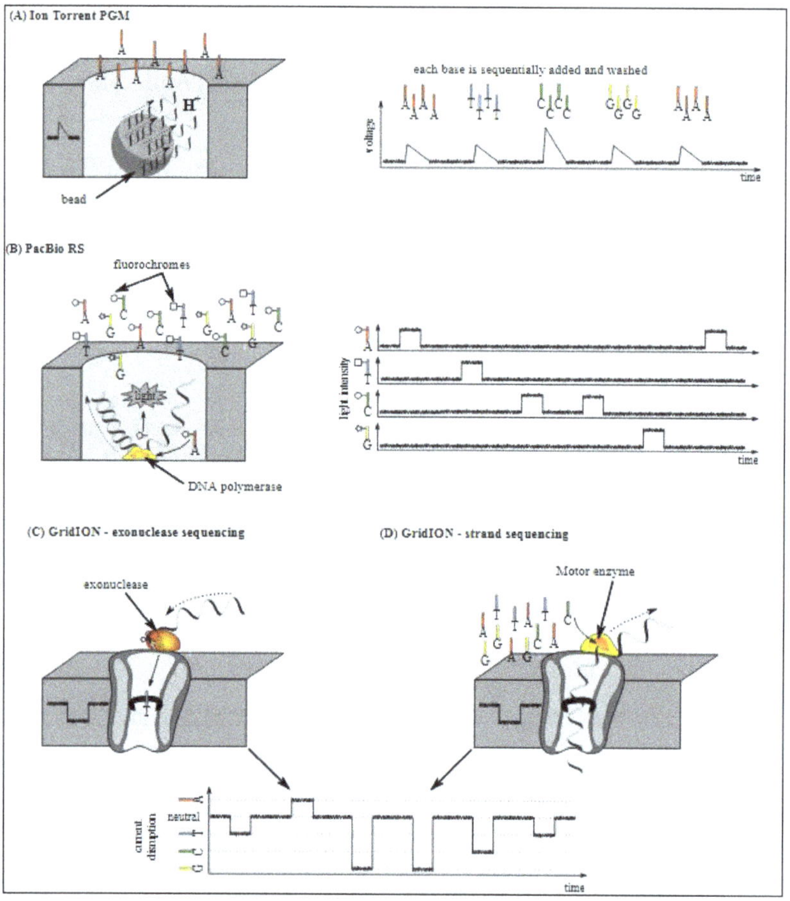

Sequencing of the TAGGCT template with IonTorrent, PacBioRS and GridION.

DNA Nanoball Sequencing

DNA nanoball sequencing is a type of high throughput sequencing technology used to determine the entire genomic sequence of an organism. The company Complete Genomics uses this technology to sequence samples submitted by independent researchers. The method uses rolling circle replication to amplify small fragments of genomic DNA into DNA nanoballs. Unchained sequencing by ligation is then used to determine the nucleotide sequence. This method of DNA sequencing allows large numbers of DNA nanoballs to be sequenced per run and at low reagent costs compared to other high-throughput sequencing platforms. However, only short sequences of DNA are determined from each DNA nanoball which makes mapping the short reads to a reference genome difficult. This technology has been used for multiple genome sequencing projects and is scheduled to be used for more.

Heliscope Single Molecule Sequencing

Heliscope sequencing is a method of single-molecule sequencing developed by Helicos Biosciences. It uses DNA fragments with added poly-A tail adapters which are attached to the flow cell surface. The next steps involve extension-based sequencing with cyclic washes of the flow cell with fluorescently labeled nucleotides (one nucleotide type at a time, as with the Sanger method). The reads are performed by the Heliscope sequencer. The reads are short, averaging 35 bp. In 2009 a human genome was sequenced using the Heliscope, however in 2012 the company went bankrupt.

Single Molecule Real Time (SMRT) Sequencing

SMRT sequencing is based on the sequencing by synthesis approach. The DNA is synthesized in zero-mode wave-guides (ZMWs) – small well-like containers with the capturing tools located at the bottom of the well. The sequencing is performed with use of unmodified polymerase (attached to the ZMW bottom) and fluorescently labelled nucleotides flowing freely in the solution. The wells are constructed in a way that only the fluorescence occurring by the bottom of the well is detected. The fluorescent label is detached from the nucleotide upon its incorporation into the DNA strand, leaving an unmodified DNA strand. According to Pacific Biosciences (PacBio), the SMRT technology developer, this methodology allows detection of nucleotide modifications (such as cytosine methylation). This happens through the observation of polymerase kinetics. This approach allows reads of 20,000 nucleotides or more, with average read lengths of 5 kilobases. In 2015, Pacific Biosciences announced the launch of a new sequencing instrument called the Sequel System, with 1 million ZMWs compared to 150,000 ZMWs in the PacBio RS II instrument. SMRT sequencing is referred to as “third-generation” or “long-read” sequencing.

Nanopore DNA Sequencing

The DNA passing through the nanopore changes its ion current. This change is dependent on the shape, size and length of the DNA sequence. Each type of the nucleotide blocks the ion flow through the pore for a different period of time. The method does not require modified nucleotides and is performed in real time. Nanopore sequencing is referred to as “third-generation” or “long-read” sequencing, along with SMRT sequencing.

Early industrial research into this method was based on a technique called ‘exonuclease sequencing’, where the readout of electrical signals occurred as nucleotides passed by alpha(α)-hemolysin pores covalently bound with cyclodextrin. However the subsequent commercial method, ‘strand sequencing’, sequenced DNA bases in an intact strand.

Two main areas of nanopore sequencing in development are solid state nanopore sequencing, and protein based nanopore sequencing. Protein nanopore sequencing utilizes membrane protein complexes such as α-hemolysin, MspA (*Mycobacterium smegmatis* Porin A) or CssG, which show great promise given their ability to distinguish between individual and groups of nucleotides. In contrast, solid-state nanopore sequencing utilizes synthetic materials such as silicon nitride and aluminum oxide and it is preferred for its superior mechanical ability and thermal and chemical stability. The fabrication method is essential for this type of sequencing given that the nanopore array can contain hundreds of pores with diameters smaller than eight nanometers.

The concept originated from the idea that single stranded DNA or RNA molecules can be electrophoretically driven in a strict linear sequence through a biological pore that can be less than eight nanometers, and can be detected given that the molecules release an ionic current while moving through the pore. The pore contains a detection region capable of recognizing different bases, with each base generating various time specific signals corresponding to the sequence of bases as they cross the pore which are then evaluated. Precise control over the DNA transport through the pore is crucial for success. Various enzymes such as exonucleases and polymerases have been used to moderate this process by positioning them near the pore's entrance.

Microfluidic Systems

There are two main microfluidic systems that are used to sequence DNA; droplet based microfluidics and digital microfluidics. Microfluidic devices solve many of the current limitations of current sequencing arrays.

Abate et al. studied the use of droplet-based microfluidic devices for DNA sequencing. These devices have the ability to form and process picoliter sized droplets at the rate of thousands per second. The devices were created from polydimethylsiloxane (PDMS) and used Forster resonance energy transfer, FRET assays to read the sequences of DNA encompassed in the droplets. Each position on the array tested for a specific 15 base sequence.

Fair et al. used digital microfluidic devices to study DNA pyrosequencing. Significant advantages include the portability of the device, reagent volume, speed of analysis, mass manufacturing abilities, and high throughput. This study provided a proof of concept showing that digital devices can be used for pyrosequencing; the study included using synthesis, which involves the extension of the enzymes and addition of labeled nucleotides.

Boles et al. also studied pyrosequencing on digital microfluidic devices. They used an electro-wetting device to create, mix, and split droplets. The sequencing uses a three-enzyme protocol and DNA templates anchored with magnetic beads. The device was tested using two protocols and resulted in 100% accuracy based on raw pyrogram levels. The advantages of these digital microfluidic devices include size, cost, and achievable levels of functional integration.

DNA sequencing research, using microfluidics, also has the ability to be applied to the sequencing of RNA, using similar droplet microfluidic techniques, such as the method, inDrops. This shows that many of these DNA sequencing techniques will be able to be applied further and be used to understand more about genomes and transcriptomes.

Methods in Development

DNA sequencing methods currently under development include reading the sequence as a DNA strand transits through nanopores (a method that is now commercial but subsequent generations such as solid-state nanopores are still in development), and microscopy-based techniques, such as atomic force microscopy or transmission electron microscopy that are used to identify the positions of individual nucleotides within long DNA fragments (>5,000 bp) by nucleotide labeling with heavier elements (e.g., halogens) for visual detection and recording. Third generation technologies aim to increase throughput and decrease the time to result and cost by eliminating the need for excessive reagents and harnessing the processivity of DNA polymerase.

Tunnelling Currents DNA Sequencing

Another approach uses measurements of the electrical tunnelling currents across single-strand DNA as it moves through a channel. Depending on its electronic structure, each base affects the tunnelling current differently, allowing differentiation between different bases.

The use of tunnelling currents has the potential to sequence orders of magnitude faster than ionic current methods and the sequencing of several DNA oligomers and micro-RNA has already been achieved.

Sequencing by Hybridization

Sequencing by hybridization is a non-enzymatic method that uses a DNA microarray. A single pool of DNA whose sequence is to be determined is fluorescently labeled and hybridized to an array containing known sequences. Strong hybridization signals from a given spot on the array identifies its sequence in the DNA being sequenced.

This method of sequencing utilizes binding characteristics of a library of short single stranded DNA molecules (oligonucleotides), also called DNA probes, to reconstruct a target DNA sequence. Non-specific hybrids are removed by washing and the target DNA is eluted. Hybrids are re-arranged such that the DNA sequence can be reconstructed. The benefit of this sequencing type is its ability to capture a large number of targets with a homogenous coverage. A large number of chemicals and starting DNA is usually required. However, with the advent of solution-based hybridization, much less equipment and chemicals are necessary.

Sequencing with Mass Spectrometry

Mass spectrometry may be used to determine DNA sequences. Matrix-assisted laser desorption ionization time-of-flight mass spectrometry, or MALDI-TOF MS, has specifically been investigated as an alternative method to gel electrophoresis for visualizing DNA fragments. With this method, DNA fragments generated by chain-termination sequencing reactions are compared by mass rather than by size. The mass of each nucleotide is different from the others and this difference is detectable by mass spectrometry. Single-nucleotide mutations in a fragment can be more easily detected with MS than by gel electrophoresis alone. MALDI-TOF MS can more easily detect differences between RNA fragments, so researchers may indirectly sequence DNA with MS-based methods by converting it to RNA first.

The higher resolution of DNA fragments permitted by MS-based methods is of special interest to researchers in forensic science, as they may wish to find single-nucleotide polymorphisms in human DNA samples to identify individuals. These samples may be highly degraded so forensic researchers often prefer mitochondrial DNA for its higher stability and applications for lineage studies. MS-based sequencing methods have been used to compare the sequences of human mitochondrial DNA from samples in a Federal Bureau of Investigation database and from bones found in mass graves of World War I soldiers.

Early chain-termination and TOF MS methods demonstrated read lengths of up to 100 base pairs. Researchers have been unable to exceed this average read size; like chain-termination sequencing alone, MS-based DNA sequencing may not be suitable for large de novo sequencing projects. Even so, a recent study did use the short sequence reads and mass spectroscopy to compare single-nucleotide polymorphisms in pathogenic *Streptococcus* strains.

Microfluidic Sanger Sequencing

In microfluidic Sanger sequencing the entire thermocycling amplification of DNA fragments as well as their separation by electrophoresis is done on a single glass wafer (approximately 10 cm in diameter) thus reducing the reagent usage as well as cost. In some instances researchers have shown that they can increase the throughput of conventional sequencing through the use of microchips. Research will still need to be done in order to make this use of technology effective.

Microscopy-based Techniques

This approach directly visualizes the sequence of DNA molecules using electron microscopy. The first identification of DNA base pairs within intact DNA molecules by enzymatically incorporating modified bases, which contain atoms of increased atomic number, direct visualization and identification of individually labeled bases within a synthetic 3,272 base-pair DNA molecule and a 7,249 base-pair viral genome has been demonstrated.

RNAP Sequencing

This method is based on use of RNA polymerase (RNAP), which is attached to a polystyrene bead. One end of DNA to be sequenced is attached to another bead, with both beads being placed in optical traps. RNAP motion during transcription brings the beads in closer and their relative distance changes, which can then be recorded at a single nucleotide resolution. The sequence is deduced based on the four readouts with lowered concentrations of each of the four nucleotide types, similarly to the Sanger method. A comparison is made between regions and sequence information is deduced by comparing the known sequence regions to the unknown sequence regions.

In Vitro Virus High-throughput Sequencing

A method has been developed to analyze full sets of protein interactions using a combination of 454 pyrosequencing and an in vitro virus mRNA display method. Specifically, this method covalently links proteins of interest to the mRNAs encoding them, then detects the mRNA pieces using reverse transcription PCRs. The mRNA may then be amplified and sequenced. The combined method was titled IVV-HiTSeq and can be performed under cell-free conditions, though its results may not be representative of in vivo conditions.

Sample Preparation

The success of any DNA sequencing protocol relies upon the DNA or RNA sample extraction and preparation from the biological material of interest.

- A successful DNA extraction will yield a DNA sample with long, non-degraded strands.
- A successful RNA extraction will yield a RNA sample that should be converted to complementary DNA (cDNA) using reverse transcriptase—a DNA polymerase that synthesizes a complementary DNA based on existing strands of RNA in a PCR-like manner. Complementary DNA can then be processed the same way as genomic DNA.

According to the sequencing technology to be used, the samples resulting from either the DNA or the RNA extraction require further preparation. For Sanger sequencing, either cloning procedures or PCR are required prior to sequencing. In the case of next-generation sequencing methods, library preparation is required before processing. Assessing the quality and quantity of nucleic acids both after extraction and after library preparation identifies degraded, fragmented, and low-purity samples and yields high-quality sequencing data.

RNA Sequencing

RNA-seq (RNA-sequencing) is a technique that can examine the quantity and sequences of RNA in a sample using next generation sequencing (NGS). It analyzes the transcriptome of gene expression patterns encoded within our RNA.

What are the Applications of RNA-seq?

RNA-seq lets us investigate and discover the transcriptome, the total cellular content of RNAs including mRNA, rRNA and tRNA. Understanding the transcriptome is key if we are to connect the information on our genome with its functional protein expression. RNA-seq can tell us which genes are turned on in a cell, what their level of expression is, and at what times they are activated or shut off[2]. This allows scientists to more deeply understand the biology of a cell and assess changes that may indicate disease. Some of the most popular techniques that use RNA-seq are transcriptional profiling, SNP identification, RNA editing and differential gene expression analysis.

This can give researchers vital information about the function of genes. For example, the transcriptome can highlight all the tissues in which a gene of unknown function is expressed, which might indicate what its role is. It also captures information about alternative splicing events, which produce different transcripts from one single gene sequence. These events would not be picked up by DNA sequencing. It can also identify post-transcriptional modifications that occur during mRNA processing such as polyadenylation and 5' capping.

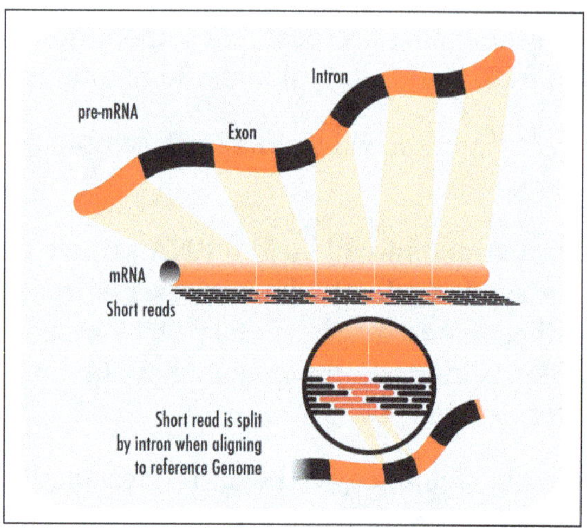

RNA-seq data uses uses short reads of mRNA which is free of intronic non-coding DNA. These reads must then be aligned back to the reference genome.

How does RNA-seq Work?

Early RNA-seq techniques used Sanger sequencing technology, a technique that although innovative at the time, was also low-throughput, costly, and inaccurate. It is only recently, with the advent and proliferation of NGS technology, have we been able to fully take advantage of RNA-seq's potential.

The first step in the technique involves converting the population of RNA to be sequenced into cDNA fragments (a cDNA library). This allows the RNA to be put into an NGS workflow. Adapters are then added to each end of the fragments. These adapters contain functional elements which permit sequencing; for example, the amplification element and the primary sequencing site. The cDNA library is then analyzed by NGS, producing short sequences which correspond to either one or both ends of the fragment. The depth to which the library is sequenced varies depending on techniques which the output data will be used for. The sequencing often follows either single-read or paired-end sequencing methods. Single-read sequencing is a cheaper and faster technique (for reference, about 1% of the cost of Sanger sequencing) that sequences the cDNA from just one end, whilst paired-end methods sequence from both ends, and are therefore more expensive and time-consuming.

A further choice must be made between strand-specific and non-strand-specific protocols. The former method means the information about which DNA strand was transcribed is retained. The value of extra information obtained from strand-specific protocols make them the favorable option.

These reads, of which there will be many millions by the end of the workflow, can then be aligned to a genome of reference and assembled to produce an RNA sequence map that spans the transcriptome.

RNA-seq vs. Microarrays: Why RNA-seq is Considered Superior?

RNA-seq is widely regarded as superior to other technologies, such as microarray hybridization. There are several reasons for RNA-seq's well-regarded status:

- Not limited to genomic sequences: Unlike hybridization-based approaches, which may require species-specific probes, RNA-seq can detect transcripts from organisms with previously undetermined genomic sequences. This makes it fundamentally superior for the detection of novel transcripts, SNPs or other alterations.
- Low background signal: The cDNA sequences used in RNA-seq can be mapped to targeted regions on the genome, which makes it easy to remove experimental noise. Furthermore, issues with cross-hybridization or sub-standard hybridization, which can plague microarray experiments, are not an issue in RNA-seq experiments.
- More quantifiable: Microarray data is only ever displayed as values relative to other signals detected on the array, whilst RNA-seq data is quantifiable.

RNA-seq also avoids the issues microarrays have in detecting very high or very low expression levels.

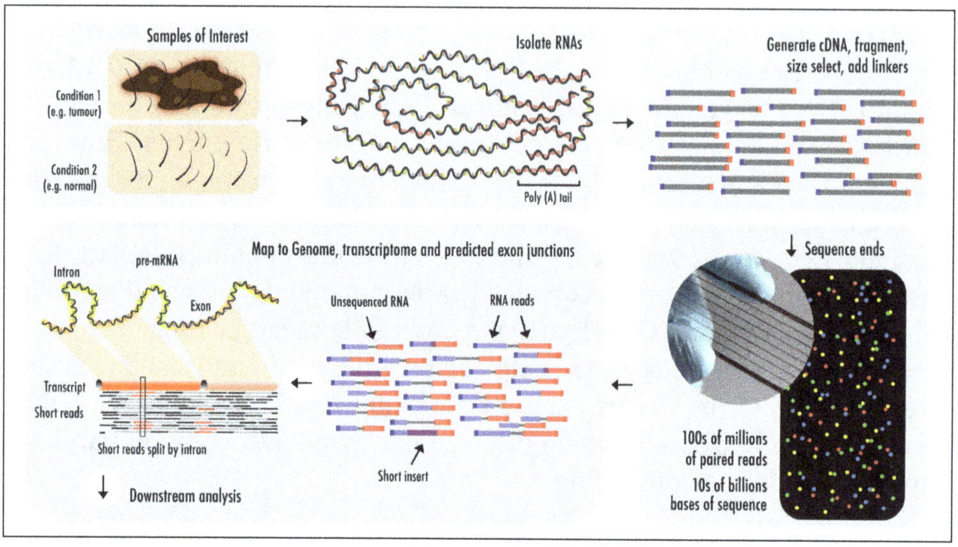

A workflow for RNA seq.

An RNA-seq Protocol Experiment Planning

Preparation prior to starting your RNA-seq experiment is essential. Questions to answer before starting include:

- What method of RNA purification are you using?
- How many reads will you need?
- Which platform will you use?
- What reference genome will you use?
- How are you assessing the quality of your RNA?
- Do you need to enrich your target RNA?
- Will you barcode your RNA?
- Have I got enough biological and technical replicates?
- Single-read or paired-end sequencing?

cDNA Library Preparation

After these points have been considered, you can start preparing your cDNA library. This will require adding the platform-specific "adapter sequences" and amplification of the DNA, but the exact procedure will be very specific to the platform used at this stage.

The amplification of the DNA involves a reverse transcriptase mediated first strand synthesis followed by a DNA polymerase-mediated second strand synthesis.

cDNA Sequencing

Once the library is prepared, and adapters added, you can use your chosen sequencing platform to sequence your cDNA library to your desired depth. Once your transcript data has been produced, you can map the data to your reference genome. The alignment process can be complicated by the presence of splice variants and modifications, and the choice of reference genome used will also vary how difficult this stage is. Software packages such as STAR are useful at this stage, as are quality control tools like Picard or Qualimap.

RNA-seq Data Analysis

After the alignment stage, you can focus on analyzing your data. Tools like Sailfish, RSEM and BitSeq will help you quantify your expression levels, whilst tools like MISO, which quantifies alternatively spliced genes, are available for more specialized analysis. There is a library of these tools out there, and reading reviews and roundups are your best way to find the right tool for your research.

References

- 1980-dna-profiling, resources: sciencelearn.org.nz, Retrieved 25 February, 2020
- Dahm R (January 2008). "Discovering DNA: Friedrich Miescher and the early years of nucleic acid research". Human Genetics. 122 (6): 565–81. Doi:10.1007/s00439-007-0433-0. PMID 17901982
- Y-chromosome-analysis: exploredna.co.uk, Retrieved 16 January, 2020
- Butler JM (2005). Forensic DNA typing : biology, technology, and genetics of STR markers (2nd ed.). Amsterdam: Elsevier Academic Press. ISBN 9780080470610. OCLC 123448124
- Applications-of-dna-profiling: mangalmay.org, Retrieved 29 March, 2020
- Chemical Synthesis, Sequencing, and Amplification of DNA (class notes on MBB/BIO 343)". Arizona State University. Archived from the original on 9 October 1997. Retrieved 29 October 2007
- DNA-Sequencing-Method-Benefits-and-Applications, Genetics-Article: biotecharticles.com, Retrieved 30 April, 2020

6

Applications of DNA and RNA

DNA and RNA find their applications in various fields such as genetic engineering, agriculture, DNA origami, DNA nanotechnology, and CRISPR and associated systems, gene expression profiling, identification of allele-specific expression, gene fusions, etc. All these diverse applications of DNA and RNA have been thoroughly discussed in this chapter.

DNA Technology in Agriculture

DNA is the main genetic materials of all cellular organisms; preserving DNA itself is one way of preserving germplasm resources. The development of biotechnology and molecular biology make it possible for us to regulate or even control the plant traits, by using DNA sequence information, such as the structure, function and mechanism etc. DNA technologies based on DNA molecular markers, transgenic technology and gene expression have been widely used in agricultural production which have showed great potential in improving agricultural yields and quality, reducing the loss that various biotic and abiotic stress caused, promoting the utilization of germplasm resource, improving breeding efficiency and strengthening the regulation of plant growth. These modern DNA technologies with high feasibility and necessity are important measures to guarantee the sustainable development of agricultural. Despite the agriculture including plant and animal production, DNA technologies in these fields share the same technical purpose and type.

There were many reports about the application of various DNA technologies in agricultural production till now; and the DNA technologies have been constantly developed and updated, which make it difficult to introduce every DNA technique.

The Application of DNA Markers in Agriculture

Since the advance of the first generation molecular marker technology restricted fragment length polymorphisms (RFLP) and the Polymerase Chain Reaction (PCR) in 1980, various DNA markers were developed and used in biology research, such as

DNA random amplified polymorphic DNA (RAPD), single nucleotide polymorphism (SNP), amplified fragment length polymorphism, (AFLP) and sequence-related amplified polymorphism (SRAP) etc. Compared with the traditional morphological markers, cytological markers, and biochemistry markers, DNA markers have much more information and higher polymorphism, and can work without the influence of plant organs, developmental stages and various environmental factors, which have been widely used in identification of crop varieties, conservation and evaluation of plant germplasm resource, analysis of genetic adversity and evolution, construction of genetic maps, cloning of important agronomic trait genes, and molecular mark assisted breeding.

Cultivar Identification and Analysis of Seed Purity

For a long time, plant cultivar identification and seed purity test were mainly based on phenotypic traits. However, the rapidly increasing crop cultivars and the close evolutional relationship among some cultivars and the confuse of the varietal names and their local names, in addition, the phenomena of 'homonym' and 'synonym' have been making it more and more difficult to collect, classify and conserve the plant resources. The incompetence of the traditional plant cultivar identification techniques calls for a more stable, simple and accurate technique. DNA molecular marker has high specificity, selectivity, simplicity, accuracy and genetic stability can reflects the differences in DNA level without environmental impacts, and thus has great advantages in seed purity identification. The DNA molecular marker technology has been used in the identification of cultivar and seed purity of rice (Oryza sativa), wheat (Triticum aestivum), maize (Zea mays), cotton (Gossypium hirsutum) and grape (Vitis Vinifera) etc.

In order to better apply the DNA markers to cultivar identification, the conversion of DNA fingerprint to the available and convenient information which can be directly used in the cultivar identification practice is crucial. However, in the early days, the lack of available technique or measure make it difficult to convert the DNA marker fingerprint to the intuitive and easy access information, which lead to the research findings on cultivar identification were barely used in the agriculture production. In the former study, the DNA marker identification results were mainly analyzed by electrophoresis band database, binary table and cluster analysis, but these information were too abstract and difficult to be used in the ordinary cultivar identification practice. Recently, with the development and application of the manual cultivar identification diagram (MCID) the cultivar identification by DNA marker has achieved the goal to serve for the agriculture directly. In MCID, the information of DNA marker primer and the polymorphic bands were marked to the cultivar identification diagram (CID) through artificial drawing, by which the cultivar identification results can be used or referred in the cultivar identification practice. Many authors applied the RAPD molecular marker to the cultivar identification of grapevine, pomegranate, loose-skin mandarin and lemon. The CID they have drawn can provide the useful information of primer and polymorphic bands which can be used to identify the cultivar of grape and pomegranate. The combination of DNA marker and the MCID make it possible to exert the advantages of DNA marker in cultivar identification sufficiently.

Purity Test.

The schematic of manual cultivar identification diagram (MCID). The plus sign (+) indicates the presence of the amplification fragment, the minus sign (–) indicates the absence of the amplification fragment, ellipsis (...) indicates that the lists of cultivars/primer/amplification fragment are not intended to be complete.

Evaluation of Germplasm Resource

Preservation and evaluation of plant germplasm resources are very important for keeping the rich diversity of breeding parents and genetic base and the rational using the germplasm resource efficiently and rationally. DNA molecular marker technology provides an important technical mean for preservation, identification, evaluation, mining and innovation of plant germplasm resources. Ram et al. and Moncada et al. analyzed the genetic diversity of field crops such as rice and grapevine by SSR marker, and this results provided important theoretical basis for the utilization of these crops. DNA molecular marker technology has become efficient technology to evaluate the germplasm resources of different crops. By using DNA molecular marker, not only can we protect the genetic integrity of germplasm resources, maintain minimum breeding population and seed amount, screen important germplasm, and preserve large amount of germplasm resources, but also we can study the genetic diversity and evolutional relation of germplasm resources. The information about their DNA level diversity and their origin and evolution relationship will greatly help us to make better use of the excellent germplasm resources of crop and provide an important scientific basis for the protection of these germplasm resources.

Heterosis Prediction

Heterosis prediction can greatly reduce the blindness and increase the pertinence of breeding, it is important to improve breeding efficiency and shorten the breeding process. By DNA marker technology, we can detect the positive markers or positive locus related to heterosis in any period or organ, and then predict the heterosis based on the heterozygosity of these locus. This technique overcome the deviation of genetic distance-based heterosis prediction (in some case, genetic distance has no relationship with heterosis). In addition, DNA marker overcome the shortcoming of isozyme-based heterosis prediction, in which it has too much restriction to be widely used. In recent years, the use of molecular markers has made some progress in the prediction of heterosis in maize, rice, cotton, rape and other crops. For example, Percy et al. found that the heterosis of boll number and weight of single cotton were significantly related the genetic distance of the molecular marker; Solomon et al. reported that the genetic distance of the molecular marker has significantly positive correlation with the heterosis of grain weight.

Construction of Genetic Map

The genetic map is the chromosome linear linkage map that uses the chromosome recombination and exchange rate as the relative length and genetic markers as the main body. It has great theoretical and practical significance in fields such as quantitative trait loci (QTL) mapping, marker-assisted selection, gene map-based cloning and comparative genomics etc. The genetic map constructed by DNA molecular marker technique give us chance to analyze the linkage relation of the genetic loci at the molecular level directly. The genetic loci we can get from the genetic map are far more than that we can get from the traditional genetic markers methods constructed based on morphology, physiology, or biochemistry. In addition, the construction cycle is shorter, and the density of markers is higher compared with the traditional ways. In recent years, with the rapid development of molecular marker technology, the genetic maps for most crops have been drawn. For example, Cui et al. constructed the high-density genetic linkage map of wheat (average genetic distance is 4.42) using DArT markers, and Blenda et al. constructed the high-density genetic linkage map of cotton (average genetic distance is 4.3) using SSR marker. Most of the field and economic crops have finished the construction of high density and even saturated genetic linkage maps, and this will be of great help to understand the information of genomic composition and structure, and will finally benefit to the breeding practice.

Gene Mapping, Cloning and Marker-assisted Breeding

Molecular marker can link to the target phenotypic traits loci, by which we can locate the important agronomic traits related genes, provide the convenience for marker-assisted breeding and map-based cloning of these genes. On the basis of precise location of the target gene by using DNA molecular markers, breeders can take advantage of the molecular markers which closely linked to or co-separated with the target genes, by which they can identify the existence of the target gene in breeding offspring, and determine whether they obtain the desired individuals. Molecular-assisted breeding makes up for the shortcomings of the traditional field selective breeding method which is time consuming, low deficiency and inaccurate. At present, molecular-assisted breeding technologies has been widely used in crop breeding and has developed a number of new crop cultivars and lines. Jena and Mackill and Shi et al. applied the molecular marker assisted breeding techniques to select the excellent crop quality traits, and succeed to select the blast resistance trait of rice and resistance to soybean mosaic virus of soybean. Gene map-based cloning technology is a technology suitable for gene cloning, which is developed on the basis of DNA molecular markers and marker genetic linkage map. Based on accurate location of the target gene by molecular markers, and use the molecular markers that are closely linked to the target gene as probes to screen genomic library, by which we can clone the large fragments of the target gene. Map-based cloning techniques has great advantages for novel gene cloning, and has been successfully used in the separation and cloning of excellent agronomic, growth and development and resistance related genes in rice and maize. For example, Tamura et al. and Gao and Lin successfully cloned salt- and insect-resistant genes in rice by map-based cloning technique, Lu et al. cloned the key enzymes genes of terpenoid metabolic pathway in maize.

Application of Gene Expression Information in Agricultural Production

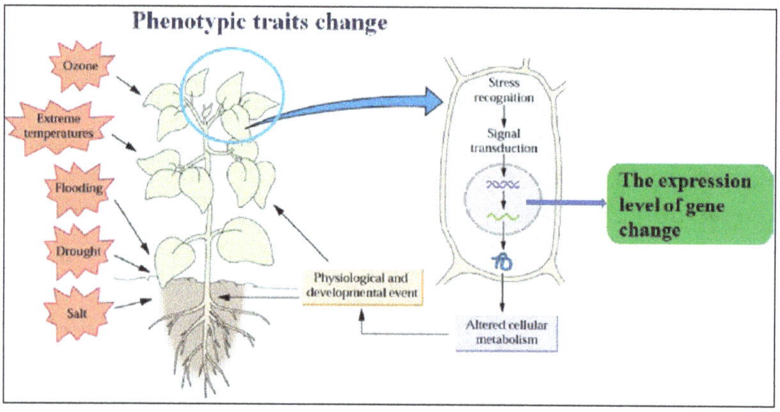

The model of gene expression information for depiction of plant growth status.

Traditionally, the growth and development condition and the phenological period of crops are the main reference basis for the agricultural production. These information are mainly collected through observing and recording the phenotypic trait of crops. This method has played important roles in the traditional agricultural production that is intuitive, simple and effective. However, the appearance of phenotypic traits always lags, and cannot feedback the real growth and development situation of crops timely and accurately. What's worse, it always too late to take remedial measures when the negative phenotype appears. Every traits of plant is genetically controlled by gene, and the gene information can accurately reflect the real growth, development and metabolic status. Furthermore, the expression of gene is always earlier than the appearance of phenotype, so monitor and diagnose the growth and development status of crops at gene level makes it possible to predict the unfavorable change or harm before it happens, and take remedial measures in advance. The development of modern molecular biology technique provides extremely rich and even redundant gene information to the researchers. Even though gene expression profile has been widely used to address the relationship between ecologically influenced or disease phenotypes and the cellular expression patterns, the information is mainly limited to laboratory experience, it is necessary to apply the crop growth and development related gene information to agricultural production, so the techniques or methods that are able to convert these gene information to convenient and easy access information are urgent.

In the era of molecular biology, there are high necessity and feasibility to apply the gene information to agricultural production. People can use genetic information to observe or monitor the growth status of crops and provide guidelines for the field management, by which we can improve the efficiency of agricultural measures such as fertilization and irrigation, and regulate the maturity and growth habits and other important growth process of crop.

Gene Information Describing the Crop Phenology

Plant phenology is the appear time of various iconic morphological characteristics such as seed germination, leaf, blossom, fruit growth and development status in plant life cycle, is the reaction of growth, development and activity of plant response to climate. By observing phenological events, we can grasp the seasonal variation of crop, guide crop planting, cultivation management and disease-pest control. However, observing and recording the phenological period need to spend a lot of time, work force and financial resources, and the data obtained is limited by the cultivar, climate, geographical environment and agricultural operations and other factors. Thus phenology and phenological period is different in different years, last year record for reference for the following year is not precise enough. Therefore, the current situation, i.e., phenological events cannot be used to accurately depict phenophases, calls for the advent of a more accurate means of depicting phenology. The expression of genes involved in plant growth and development is affected by the common regulation of plant growth and environmental conditions, so the researchers can more accurately describe the phenological period of crop growth by using the gene information, i.e., gene phenology, and then according to gene phenology we can early elaborated plans and employed appropriate management.

When the changes of external environment caused the advance or delay of phenology, the expression of related genes at the molecular level invariably precedes the emergence of phenological phenomenon, therefore, we can timely and reasonable judge and implement farming operation. Wang et al., found changes in expression information of flower development related gene APETALA1 and APETALA3 in grape can be used to describe the phenology of grapes, and can guide the precise fertilization in grape production, restrain flower and fruit drop and promote the development of flower and fruit.

As more and more genes related to crop growth and development genes have been excavated, the researchers were able to describe different phenological events in different crops at different developmental stages systematically, precisely and quickly. This will enable the use of gene information to describe phenology and take appropriate management measures become simpler, more efficient and cost less. This strategy promotes the practice of using gene information in agricultural production.

Prediction of Crop Growth Status under Stress

Gene expression information can also be used to predict the effects of environmental stress on the growth and development of crops, which is convenient for people to take appropriate management measures. Crops sometimes suffer various biotic or abiotic stresses in agricultural production, resulting in reduction in agronomical yield, falling off in quality, and will affect agricultural output.

In the past, people grasp the growth state under stress through the phenotypic traits of crop. Nevertheless, these phenotypic traits require a certain amount of time, and the effect of some stress on crop is difficult to observe from the outside. More often, when people have observed the phenotypic traits of crop under stress and then take measures, it is often difficult to bring the plant back to normal state. Therefore, it is not the suitable way and measure to guide the remedial or recovery of crops under stress through the phenotype observation of the growth state of crop under adversity.

There are many genes of crop involved in transcriptional response to stress such as drought, high salinity or low temperature, the expression level of these genes increased or decreased affected by the stress condition. Therefore, through the expression information of these resistance related genes, researchers will be able to know whether the crop encounter unfavorable growth conditions, and take timely remedial measures.

Krasensky and Jonak studied the expression patterns of stress-inducible genes in drought, extreme temperature and salinity stress. The beginning, end and changes in the expression of genes related to plant signaling events involved in responses to environmental stresses such as salicylic acid, jasmonic acid and abscisic acid and calcium ion, reflect the growth state of crop under stress. Researchs about the expression information of gene in different tissues and different developmental stages under different environmental conditions will provide support for the prediction of growth state of crop under stress at the molecular level. These researches not only contribute to complementing understanding of metabolic adjustments of plant under stress, but also play an important role in increasing agricultural production, reducing the consumption of resources.

Evaluation of the Effect of Fertilizers

In the past, determining fertilization time, fertilization types, and the amount of fertilizers in agricultural production are general implemented based on crop growth and phenology, generally is experience operation, and would be greatly affected by geographical and environmental conditions, thus it is undoubtedly difficult to accurately guide the effect of fertilization. These practices not only failed to achieve the best benefits of fertilizer application, but also resulted in the waste of fertilizer and environmental pollution. Therefore, how to invest a reasonable input of nutrient, at the same time guarantee the quality and yield of crops, and how to choose the more effective fertilizers and more scientific fertilizing methods, are issues of concern.

Modern DNA technology makes it possible for people to understand the effect of fertilization promoting plant growth and development at the molecular level. Chu et al. found we can determine the effect of fertilization by the expression of some genes related to nitrogen metabolism in grape. Through the analysis of expression level of some gene such as glutamine synthetase (GS) and glutamate dehydrogenase (GDH), the time of these gene reached the highest expression level was 6h earlier after foliar fertilization compared with soil fertilization, and gene expression levels were higher, these results partly revealed the reason why foliar fertilization take effect quickly. The use of genetic information not only can accurately clarify the influence of conventional fertilization on crop growth and development; also can more accurately assess the use efficiency of fertilizer, the work speed of different fertilizing methods such as foliar and soil fertilizer, the fertility of different fertilizer and so on. Gene expression information can reflect the metabolism and transport of fertilizer in crop, so through analysis of spatio-temporal expression of related gene, we can obtain the gene information on the response of crops in different kinds and concentration of fertilizer in different growth periods. On this basis, combined with the status of crop growth and development, we can establish more scientific and reasonable fertilization measures in different growth stages, such as the period of fertilization, the type of fertilization and the amount of fertilizer. These measures will greatly save fertilizer dosage, reduced investment in agriculture, increased the income of farmers, and reduced the environmental pollution.

Crop cultivation along with many farming operations, each farming operation will cause the corresponding changes in gene expression at the DNA level, the usage of DNA technology to detect such changes at the molecular level can help us to understand regulation mechanism of farming operation on crop growth and development. Now there were more and more studies about the influence of various cultivation operations on the expression level of related genes. The expression level of some genes of grape increased and expressed early because of treatment with abscisic acid, similarly, branch girdling makes the high expression level of citrus genes about 1 week in advance.

According to soil characteristics and the needs of crop growth development, we can use gene information to manage every growth stage of crop, and decide the consumption of various agricultural material (fertilizer, herbicide, pesticide, and hormone etc.), and which cultivation measures (girdling, bagging etc.) to use. Thus, we can make full use of the potential of soil and crop, archive the best results from field management technology, meet the needs of crop growth, and reduce agricultural materials inputs, thereby reduced the material consumption, increased commercial profits, protected the ecological environment, and realized the sustainable development of agriculture.

In total, plant growth and development are controlled by the programmed (by time, tissue and abundance) expression of suites of genes in response to exogeneous or endogeneous queues. Hence, there is a need to generate genome-wide expression data from a range of tissues/developmental stages in order to understand and relate phenotypic traits and gene expression profiles. And data at the transcriptome and epigenome level can contribute greatly to the application of gene information in agriculture.

Applications of Transgenic Technology in Agriculture

Transgenic technology can improve accurately crop quality and yield, compared with traditional crop, the all traits like yield, stress resistance (including disease-resistant, insect-resistant, cold-resistant and herbicide-resistant) and nutritional quality of genetically modified crops will be significantly heighten. The commercialization of genetically modified crops dramatically lowering the cost of agriculture, and bring continuously a great deal of benefits of economic, environmental and social in the world.

Prevention and Control of Agricultural Disease and Pest

Transgenic technology is widely used to control the agricultural disease and insect-pest, reduce the cost of agricultural production, improve the effect of disease and insect control, reduce pesticide pollution, and increase crop production, thus brought new vigor and hope in the prevention and control of crop diseases and pest.

The research and promotion of transgenic herbicide-resistant crop has been dominant in the transgenic crop, to date there are many herbicide-resistant cultivars were created by transgenic technology in cotton, maize, soybeans and rape. Transgenic insect-resistant crop is one of the most important kinds of transgenic crops, in the past 10 years, transgenic insect-resistant crop such as cotton, maize and potato have achieved fruitful results. In recent years, plant disease-resistance genetic engineering has achieved many breakthroughs, and has bred many kinds of transgenic disease-resistant crop such as the Fusarium wilt- and Verticillium wilt-resistance transgenic cotton.

Currently, insect resistance, disease resistance, herbicide-resistant transgenic maize, cotton, potato and soybean have entered the commercialization stage, and have a larger cultivation area.

Quality Improvement

Transgenic technique can be applied to improve the qualities of crop, including protein content, amino acid composition, starch composition, polysaccharide compounds and lipid in seeds and other storage organs (tubers, tuberous root, bulbs, etc.), finally improve food quality, increase nutrition and ameliorate health function and processing performance etc.

Enhanced Resistance

Transgenic technology is also widely applied in breeding new cultivars possessed high drought-, salt- and cold-resistance. The rice which introduced into c-repeat binding factor (CBF) possess significantly resistance to abiotic stress. The offspring of rice introduced into Arabidopsis thaliana dehydration responsive element binding (DREB) gene exhibited better growth status than non-transgenic plants under drought stress. The tobacco introduced into S-adenosylmethionine decarboxylase (SAMDC) gene grew well under abiotic stress. A bacterial mannitol-1-phosphate dehydrogenase (mtlD) gene was transferred into tomato which showed a significant resistance to the low temperature stress.

Transgenic crops have been extended into many aspects, such as biotic and abiotic stress-tolerant, fertility, yield and quality improvement furthermore, some techniques have been or are being applied in crop breeding program. Golden rice and drought-tolerant maize created early by transgenic technique could be used in actual production cultivation.

Transgenic cotton is among the first transgenic plants commercially adopted around the world. Transgenic cotton has offered many important environmental, social, and economic benefits, including reduced usage of pesticides, indirect increase of yield, minimizing environmental pollution, and reducing labor and cost. Although transgenic cotton plants with disease-resistance, abiotic stress tolerance, and improved fiber quality have been developed in the past decades, insect-resistant and herbicide-tolerant cotton are the two dominant transgenic cottons in the transgenic cotton market. Thus in the future, the development of transgenic crops no doubt will focus on fields like stress resistance, quality improvement, yield increased, functional components, this will bring opportunities and challenges for the development of world agriculture.

Molecular Design Breeding

In recent years, molecular design breeding developed on basis of the combination of the DNA technology and cultivar improvement has gradually attracted people's attention. Before the field experiments, molecular design breeding would integrate the information related breeding program on the computer, and simulate the implementation plan, thus considered more factors, more comprehensively, and put forward the best strategies of selection of parental and progeny, to enhance predictability in the breeding process, thereby substantially improve breeding efficiency.

Molecular design breeding will greatly improve the theory and technology of plant breeding, and make the traditional breeding technology more efficiently and targeted. At present, molecular design breeding have achieved certain effects in wheat, rice, soybean and rape. Molecular design breeding is a new and highly integrated research field; we need to use multiple DNA techniques such as mapping and analysis of agronomic QTL, the mapping and cloning of important trait related genes, construction of genetic linkage map and prediction of heterosis. Among them, molecular marker-assisted breeding and gene genetic transformation techniques are two of the most basic and most important technologies in molecular design breeding, which are also key factors to restrict the development of molecular design breeding. With the innovation and breakthrough of both technologies in the further, the development of crop molecular design breeding will be more rapid.

In the development of future, molecular design breeding is bound to become an integrated breeding method that would combine multiple disciplines and would be constantly improved. Molecular design breeding will in the future have a profound influence on the development of theory and technology of crop breeding, and DNA technology is playing an irreplaceable role in this process.

Gel Electrophoresis

Gel electrophoresis is a commonly used technique for resolving proteins or nucleic acids. In general, a sample of one particular type of macromolecule (protein, DNA, or RNA) is placed in a well at or near the end of a gel matrix (gel). The composition of an electrophoresis gel is a semisolid open meshwork of interlinked linear strands. A gel is cast as a thin slab with a number of sample wells. After the wells of a gel are loaded with sample, an electric field is applied across the gel, and charged macromolecules of the same size are driven together in the direction of the anode through the gel as discrete invisible bands of material. The distance that a band moves into a gel depends on the mass of its macromolecules and the size of the openings (pore size) of the gel. The smaller macromolecules travel further than the larger ones. The progress of gel electrophoresis is monitored by observing the migration of a visible dye (tracking dye) through the gel. The tracking dye is a charged, low-molecular-weight compound that is loaded into each sample well at the start of a run. When the tracking dye reaches the end of the gel, the run is terminated. The bands, which are aligned in a lane under each well, are visualized by staining the gel with a dye that is specific for protein, DNA, or RNA. Discrete bands are observed when there is enough material present in a band to bind the dye to make the band visible and when the individual macromolecules of a sample have distinctly different sizes. Otherwise, a band is not detected.

If there is little or no difference among the sizes of the macromolecules in a concentrated sample, a smear of stained material is observed. The intensity of a stained band reflects the frequency of occurrence of a macromolecule in a sample. The molecular mass (molecular weight) of a gel-fractionated macromolecule (band) is determined from a standard curve that is based on a set of macromolecules of known molecular mass (size markers) that covers the separation range of the gel system and is run in one or both of the outside lanes (calibrator lanes) of the same gel as the samples. The logarithm of the molecular mass of a size marker is related to its relative mobility (Rf) through a gel. The value of Rf is defined as the distance traveled by a band divided by the distance traveled by the tracking dye (ion front). The relationship between the logarithm of the molecular mass of each size marker and its Rf value is plotted. Then, with this standard curve, a molecular mass is calculated for each band in a lane. The units of molecular mass for proteins and double-stranded and single-stranded nucleic acids are daltons, base pairs, and bases, respectively. The size markers are included in the same gel as the samples because the extent of mobility of a macromolecule(s) varies from one electrophoretic run to the next. Polyacrylamide is the preferred gel system for separating proteins. Copolymerization of monomeric acrylamide and the cross-linker bisacrylamide forms a lattice of crosslinked, linear polyacrylamide strands. The pore size of a polyacrylamide gel is determined by the concentration of acrylamide and the ratio of acrylamide to bisacrylamide. For many applications, a protein sample is treated with the anionic detergent sodium dodecyl sulfate (SDS) before electrophoresis.

The SDS binds to proteins and dissociates most multichain proteins. Each SDS-coated protein chain has a similar charge-to-mass ratio. Consequently, during electrophoresis, the separation of the SDS– protein chains is based primarily on size, and the effect of conformation is eliminated. SDS–polyacrylamide gel electrophoresis with a 10% polyacrylamide gel resolves proteins that range from 20 to 200 kilodaltons (kDa). Agarose, which is a polysaccharide from seaweed, is used routinely as the gel matrix for the electrophoretic separation of medium-size nucleic acid molecules. A 1.0% agarose gel can resolve duplex DNA chains that range from about 600 to 10,000 bp. Specialized agarose gel electrophoresis systems are available for fractionating DNA molecules with millions of base pairs, denatured DNA, and denatured RNA. In addition, for specific purposes, polyacrylamide gels are used for separating DNA molecules. For example, DNA chains that are as small as 6 bases and that differ from each other by 1 nucleotide can be resolved with a 20% polyacrylamide gel.

Restriction Endonucleases

The chief progression in the arena of recombinant DNA technology is the identification and confirmation of enzymes which are efficient in specific cleavage of duplex DNA into distinct reproducible fragments. Many of the prokaryotes are confirmed to be the resource where restriction enzymes can be identified. Restriction enzymes specifically recognize and cleave both strands of DNA at a definite base pair sequence. An axis of symmetry exists at most of the target cleavage sites. There can be a symmetrical cleavage occurring at the axis on either side thereby yielding overlying 5' or 3' single-stranded termini. The naming of restriction endonucleases is carried out as per the standard nomenclature. The first three letters represent the genus and species of the host organism. It is followed by a strain identification or a letter which indicates if the enzyme is encoded by an extrachromosomal bacteriophage or plasmid. If there exists more than one restriction endonuclease in the same host, it is denoted by a Roman numeral.

In addition to EcoRI, more than 3,700 type II restriction endonucleases with about 250 different recognition sites have been isolated from various bacteria. The naming protocol for these enzymes is the same as that for EcoRI; the genus is the capitalized letter, and the first two letters of the species name are in lowercase letters. The strain designation is occasionallyadded to the name, such as R in EcoRI, or the serotype of the source bacteriumis sometimes noted, such as d in HindIII. The Roman numerals are used to designate the order of characterization of different restriction endonucleases from the same organism. For example, HpaI and HpaII are the first and second type II restriction endonucleases that were isolated from Haemophilus parainfluenzae.

The palindromic sequences where most type II restriction endonucleases bind and cut a DNA molecule are within the recognition sites. Some restriction endonucleases digest (cleave) DNA, leaving 5′ phosphate extensions (protruding ends, or sticky ends) with recessed 3′ hydroxyl ends; some leave 3′ hydroxyl extensions with recessed 5′ phosphate ends; and some cut the backbones of both strands within a recognition site to produce blunt-ended (flush-ended) DNA molecules. The lengths of the recognition sites for different enzymes can be four, five, six, eight, or more nucleotide pairs. Because of the frequency with which their recognition sites occur in DNA, restriction endonucleases that cleave within sites of four (four-cutters) and six (six-cutters) nucleotide pairs are used for most of the common molecular-cloning protocols.

In addition, ready access to a variety of restriction endonucleases adds versatility to gene-cloning strategies. Type IIS restriction endonucleases form a subgroup of the type II category of restriction enzymes and are occasionally used for cloning and other molecular studies, such as multiplex polony sequencing. These enzymes have the fascinating feature of cutting DNA, usually in both strands, a fixed number of nucleotides away from one end of the recognition site. Moreover, any particular sequence of nucleotides may be present between the binding sequence and the cut sites. The cleavages for most type IIS restriction enzymes are staggered. For example, the FokI restriction endonuclease binds to GGATG CCTAC and cuts 9 nucleotidesdownstream on the upper strand and 13 nucleotides downstream on the lower strand, producing a recessed 3′hydroxyl end and a 4-nucleotide extension at the 5′phosphate end.

Gene Cloning

To successfully fulfil most of the aim via recombinant DNA technology, there is a requisite for the isolation of individual genes from genomic DNA of reference host. Major four steps are involved in the process of gene cloning and since there are multiple numbers of methods for each step, an investigator can opt one out of many to perform the cloning. To describe in brief, gene cloning is achieved by the restriction endonucleases mediated digesting of host DNA followed by the fusion of desired fragments into any one of four basic vector systems. The vector, in fact, is introduced into an alternative host usually bacteria in which the genes are propagated successfully. Plasmids are known to be one of the most suitable propagating vectors for DNA fragments. These cloning vehicles are extrachromosomally replicating double-stranded DNA molecules which are circular in form, and reside in bacteria. Plasmid vectors used in laboratory scale possess common features such as the origin of replication, the potency to exist in high copy number per bacterial cell, phenotype conferring gene (eg. antibiotic resistance) that permits the selection of plasmid incorporated host bacteria.

The vector and the DNA both that is required to be cloned are processed with a similar restriction endonuclease, producing molecules with well-matched ends. Commercially available plasmid vectors harbour polylinker sequences with a number of cleavage sites of restriction endonuclease and thereby these vectors will suitably receive DNA produced by digestion with multiple enzymes. The hanging end of the vector as well as the insert are capable to join one another via complementary base pairing. The DNA ligase enzyme produces covalent bonds amongst DNA strands at these positions giving rise to a closed circular recombinant plasmid. The competent bacterial cells and the plasmid are then incubated together so that the cells are able to uptake the exogenous DNA possible by multiple methods. The bacterial cells that have taken up the plasmid can be subjected to selection via plating on an agar plate having an antibiotic. The wild type bacterial host is generally vulnerable to the antibiotic and thereby only those cells that have the plasmid obtain the gene that coding for antibiotic resistance and hence are able to grow on the agar media. This type of selection distinguishes only bacteria that are able to uptake circular plasmids where as it does not differentiate plasmids inserted with recombinant DNA from recircularized vector DNA missing inserts.

Role of DNA in Genetic Engineering

Medicine

Genetic engineering has resulted in a series of medical products. The first two commercially prepared products from recombinant DNA technology were insulin and human growth hormone, both of which were cultured in the E. coli bacteria. Since then a plethora of products have appeared on the market, including the following abbreviated list, all made in E. coli:

- Tumor necrosis factor: Treatment for certain tumor cells.
- Interleukin-2 (IL-2): Cancer treatment, immune deficiency, and HIV infection treatment.
- Prourokinase: Treatment for heart attacks.

- Taxol: Treatment for ovarian cancer.
- Interferon: Treatment for cancer and viral infections.

In addition, a number of vaccines are now commercially prepared from recombinant hosts. At one time vaccines were made by denaturing the disease and then injecting it into humans with the hope that it would activate their immune system to fight future intrusions by that invader. Unfortunately, the patient sometimes still ended up with the disease.

With DNA technology, only the identifiable outside shell of the microorganism is needed, copied, and injected into a harmless host to create the vaccine. This method is likely to be much safer because the actual disease-causing microbe is not transferred to the host. The immune system is activated by specific proteins on the surface of the microorganism -e. DNA technology takes that into account and only utilizes identifying surface features for the vaccine. Currently vaccines for the hepatitis B virus, herpes type 2 viruses, and malaria is in development for trial use in the near future.

Agriculture

Crop plants have been and continue to be the focus of biotechnology as efforts are made to improve yield and profitability by improving crop resistance to insects and certain herbicides and delaying ripening (for better transport and spoilage resistance). The creation of a transgenic plant, one that has received genes from another organism, proved more difficult than animals. Unlike animals, finding a vector for plants proved to be difficult until the isolation of the Ti plasmid, harvested from a tumor-inducing (Ti) bacteria found in the soil. The plasmid is "shot" into a cell, where the plasmid readily attaches to the plant's DNA. Although successful in fruits and vegetables, the Ti plasmid has generated limited success in grain crops.

Creating a crop that is resistant to a specific herbicide proved to be a success because the herbicide eliminated weed competition from the crop plant. Researchers discovered herbicide-resistant bacteria, isolated the genes responsible for the condition, and "shot" them into a crop plant, which then proved to be resistant to that herbicide. Similarly, insect-resistant plants are becoming available as researchers discover bacterial enzymes that destroy or immobilize unwanted herbivores, and others that increase nitrogen fixation in the soil for use by plants.

Geneticists are on the threshold of a major agricultural breakthrough. All plants need nitrogen to grow. In fact, nitrogen is one of the three most important nutrients a plant requires. Although the atmosphere is approximately 78 percent nitrogen, it is in a form that is unusable to plants. However, a naturally occurring rhizobium bacterium is found in the soil and converts atmospheric nitrogen into a form usable by plants. These nitrogen-fixing bacteria are also found naturally occurring in the legumes of certain plants such as soybeans and peanuts. Because they contain these unusual bacteria, they can grow in nitrogen-deficient soil that prohibits the growth of other crop plants. Researchers hope that by isolating these bacteria, they can identify the DNA segment that codes for nitrogen fixation, remove the segment, and insert it into the DNA of a profitable cash crop! In so doing, the new transgenic crop plants could live in new fringe territories, which are areas normally not suitable for their growth, and grow in current locations without the addition of costly fertilizers.

Animal Husbandry

Neither the use of animal vaccines nor adding bovine growth hormones to cows to dramatically increase milk production can match the real excitement in animal husbandry: transgenic animals and clones.

Transgenic animals model advancements in DNA technology in their development. The mechanism for creating one can be described in three steps:

- Healthy egg cells are removed from a female of the host animal and fertilized in the laboratory.
- The desired gene from another species is identified, isolated, and cloned.
- The cloned genes are injected directly into the eggs, which are then surgically implanted in the host female, where the embryo undergoes a normal development process.

It is hoped that this process will provide a cheap and rapid means of generating desired enzymes, other proteins, and increased production of meat, wool, and other animal products through common, natural functions.

Ever since 1997 when Dolly was cloned, research and experimentation to clone useful livestock has continued unceasingly. The attractiveness of cloning is the knowledge that the offspring will be genetically identical to the parent as in asexual reproduction. Four steps describe the general process:

- A differentiated cell, one that has become specialized during development, with its diploid nucleus is removed from an animal to provide the DNA source for the clone.
- An egg cell from a similar animal is recovered and the nucleus is removed, leaving only the cytoplasm and cytoplasm organelles.
- The two egg cells are fused with an electric current to form a single diploid cell, which then begins normal cell division.
- The developing embryo is placed in a surrogate mother, who then undergoes a normal pregnancy.

Vectors

Plasmids are self-replicating, double-stranded, circular DNA molecules that are maintained in bacteria as independent extrachromosomal entities. Virtually all bacterial genera have natural plasmids. Some plasmids carry information for their own transfer from one cell to another (e.g., F plasmids), others encode resistance to antibiotics (R plasmids), others carry specific sets of genes for the utilization of unusual metabolites (degradative plasmids), and some have no apparent functional coding genes (cryptic plasmids). Although they are not typically essential for bacterial cell survival under laboratory conditions, plasmids often carry genes that are advantageous under particular conditions. Plasmids can range in size from less than 1 kb to more than 500 kb. Each plasmid has a sequence that functions as an origin of DNA replication; without this site, it cannot replicate in a host cell.

Some plasmids are represented by 10 to 100 copies per host cell; these are called high-copy-number plasmids. Others maintain one to four copies per cell and are called low-copy-number plasmids. Seldom does the population of plasmids in a bacterium make up more than approximately 0.1 to 5.0% of the total DNA. When two or more different plasmids cannot coexist in the same host cell, they are said to belong to the same incompatibility group, but plasmids from different incompatibility groups can be maintained together in the same cell. This coexistence is independent of the copy numbers of the individual plasmids. Some microorganisms have been found to contain as many as 8 to 10 different plasmids. In these instances, each plasmid can carry out different functions and have its own unique copy number, and each belongs to a different incompatibility group. Some plasmids, because of the specificity of their origin of replication, can replicate in only one species of host cell. Other plasmids have less specific origins of replication and can replicate in a number of bacterial species. These plasmids are called narrow and broad-host-range plasmids, respectively. As autonomous, self-replicating genetic elements, plasmids have the basic attributes to make them potential vectors for carrying cloned DNA. However, naturally occurring (unmodified, or non-engineered) plasmids often lack several important features that are required for a high-quality cloning vector.

The more important features are:

- A choice of unique (single) restriction endonuclease recognition sites into which the insert DNA can be cloned.
- One or more selectable genetic markers for identifying recipient cells that carry the cloning vector–insert DNA construct. In other words, plasmid cloning vectors have to be genetically engineered.

An assortment of vector systems is now accessible to meet the demands of a specific experimental procedure. Individually each has precise pros and cons. Bacteriophage vectors have served a predominantly vital role in molecular biology. Bacteriophages are viruses with a liking towards bacteria whose genome (prototype A-bacteriophage vector) is a double-stranded, linear DNA molecule having 50 kilobases (kb) length. Each end is made up of 12 bp complementary DNA that is single-stranded in nature. Once taken up by the bacteria the virus undertakes a circular form at the time when the termini are combined and sealed by the action of host ligase. Two mechanisms of replication are likely to occur. In one of the cases, the circular DNA is multiplied into a high copy number and the host cell ultimately experiences lysis releasing the infectious viral particles. Otherwise, the bacteriophage DNA may integrate with the host bacterial DNA commonly named as the lysogenic pathway. In this scenario the viral DNA remains dormant, expressing only a handful of genes. Bacteriophages have been adapted to receive and proliferate exogenous DNA. The selection of a specific vector is dependent on the fragment size that is required to be cloned, the restriction enzyme applied in the generation of the fragment and if the cloned fragment is required to express into protein in the bacterial host. In this last context, numerous expression vectors permit the cloning of foreign inserts following the location of a strong bacterial promoter required to let the proficient transcription in the bacterial host.

Recombinant DNA Technology in Medicine

Recombinant DNA technology is the technique of genetic engineering in which recombinant DNA is prepared by cutting the DNA into small fragments and joining different fragments together taken from different organisms. This technique makes it possible to take any gene from any specie and place this gene in any other organism or specie. It is similar to cloning because when the foreign gene is incorporated in an organism like bacteria then multiple copies are made through cloning to use the gene in different applications.

- Applications in Medicine: Recombinant DNA technology had made it possible to treat different diseases by inserting new genes in place of damaged and diseased genes in the human body. It has brought many revolutionary changes in the field of medicine and introduced such methods of treating diseases and delivering the drug which were just imaginary.

- Insulin: Insulin is a hormone made up of protein. It is secreted in the pancreas by some cells called as islet cells. This hormone is responsible for controlling the glucose level in humans. If a person has decreased amount of insulin in his body, he will suffer from a disease called diabetes. Recombinant DNA technology has allowed the scientists to develop human insulin by using the bacteria as a host cell and it is also available in the market. It is believed that the drugs produced through microbes are safer than the drugs produced traditionally.

- Vaccines: Vaccine is a biological substance which is prepared to from the suspension of weak or dead pathogenic cells. It is injected in the body to enhance the production of antibodies against particular antigen. Recombinant DNA technology enables the scientists to develop vaccines by cloning the gene used for protective antigen protein. Viral vaccines are most commonly developed through this technology for example, Herpes, Influenza, Hepatitis and Foot and Mouth Disease.

- Human Growth Hormones: Human growth hormone is a polypeptide hormone. It is responsible for growth, reproduction of the cells and regeneration in humans as well as animals. It is secreted by somatotroph cells present in the pituitary glands. In recent years, scientists have developed many growth hormones using recombinant DNA technology. The disease of dwarfism is treated with this hormone.

- Monoclonal Antibodies: When a foreign object enters the body, immune system of the body releases a specific protein called as antibody. Hybridoma technology has made it possible to produce monoclonal antibodies. In this technique, the lymphocytes or B cells are joined with myeloma cells; the resulting substance is called as Hybridoma. This Hybridoma produces unlimited antibodies in the culture. The antibody produced is called as monoclonal antibody. These antibodies are used to produce vaccines against different viral infections.

- Interferon: A glycoprotein which has the ability to block the multiplication or division of viruses in the cells or in the nearby cells is called as interferon. Interferon can be used to treat cancer like hairy cell leukemia. Recombinant DNA technology produces this protein using E. coli. Interferon alpha is used to treat lymphoma and myelogenous leukemia.

- Antibiotics: Antibiotics are the chemical substances which are used against bacterial infections. They can be produced by microorganisms as well as in the laboratory. They have the ability to destroy bacteria or other harmful microbes which cause infections in the body. Alexander Fleming discovered penicillin for the first time in 1928 using recombinant DNA technology. Other biotechnological techniques are also being used to produce antibiotics.

- Infectious Diseases: Many diseases are diagnosed by conducting certain tests. Recombinant DNA technology has allowed the development of many tests which are being used to diagnose diseases like TB and cancer. Other diseases like measles, small pox and hepatitis are also diagnosed through tests and if they are not diagnosed properly, they can be a threat to human health. In the diagnosis process, certain pathogens are isolated and identified, and then diagnostic kits are produced when the genome of the specific pathogen is known to kill it or block its pathogenic activity.

DNA Nanotechnology

The DNA structure at left (schematic shown) will self-assemble into the structure visualized by atomic force microscopy at right. DNA nanotechnology is the field which seeks to design nanoscale structures using the molecular recognition properties of DNA molecules.

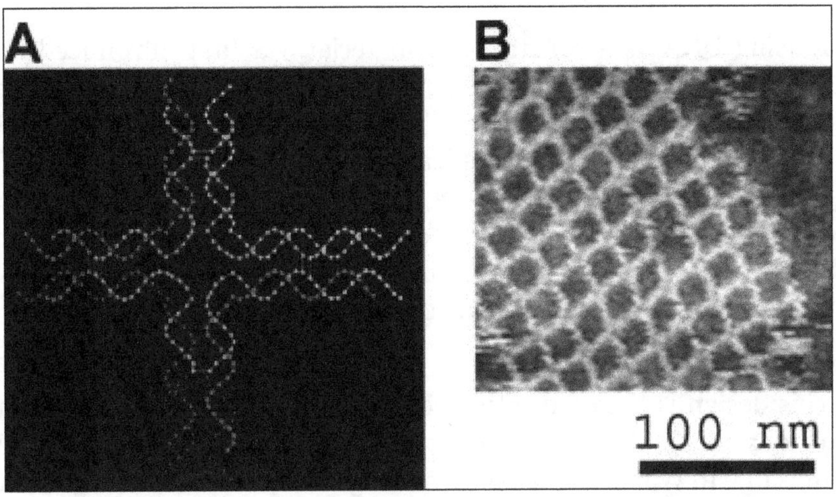

DNA nanotechnology uses the unique molecular recognition properties of DNA and other nucleic acids to create self-assembing branched DNA complexes with useful properties. DNA is thus used as a structural material rather than as a carrier of biological information. This has led to the creation of two-dimensional periodic lattices (both tile-based as well as using the "DNA origami" method) as well as three-dimensional structures in the shapes of polyhedra. Nanomechanical devices and algorithmic self-assembly have also been demonstrated, and these DNA structures have been used to template the arrangement of other molecules such as gold nanoparticles and streptavidin proteins.

DNA Origami

DNA origami is the nanoscale folding of DNA to create non-arbitrary two- and three-dimensional shapes at the nanoscale. The specificity of the interactions between complementary base pairs make DNA a useful construction material, through design of its base sequences. DNA is a well-understood material that is suitable for creating scaffolds that hold other molecules in place or to create structures all on its own.

DNA origami was the cover story of Nature on March 16, 2006. Since then, DNA origami has progressed past an art form and has found a number of applications from drug delivery systems to uses as circuitry in plasmonic devices; however, most applications remain in a concept or testing phase.

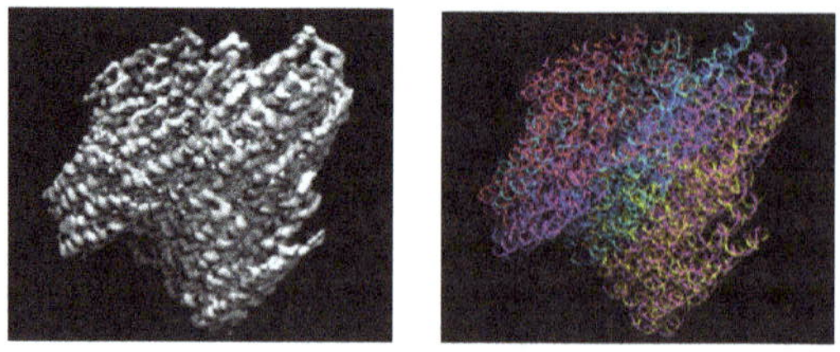

DNA origami object from viral DNA visualized by electron tomography.
The map is at the and atomic model of the DNA colored.

The idea of using DNA as a construction material was first introduced in the early 1980s by Nadrian Seeman. The current method of DNA origami was developed by Paul Rothemund at the California Institute of Technology. The process involves the folding of a long single strand of viral DNA (typically the 7,249 bp genomic DNA of M13 bacteriophage) aided by multiple smaller "staple" strands. These shorter strands bind the longer in various places, resulting in the formation of a pre-defined two- or three-dimensional shape. Examples include a smiley face and a coarse map of China and the Americas, along with many three-dimensional structures such as cubes.

To produce a desired shape, images are drawn with a raster fill of a single long DNA molecule. This design is then fed into a computer program that calculates the placement of individual staple strands. Each staple binds to a specific region of the DNA template, and thus due to Watson-Crick base pairing, the necessary sequences of all staple strands are known and displayed. The DNA is mixed, then heated and cooled. As the DNA cools, the various staples pull the long strand into the desired shape. Designs are directly observable via several methods, including electron microscopy, atomic force microscopy, or fluorescence microscopy when DNA is coupled to fluorescent materials.

Bottom-up self-assembly methods are considered promising alternatives that offer cheap, parallel synthesis of nanostructures under relatively mild conditions.

Since the creation of this method, software was developed to assist the process using CAD software. This allows researchers to use a computer to determine the way to create the correct staples needed to form a certain shape. One such software called caDNAno is an open source software for creating such structures from DNA. The use of software has not only increased the ease of the process but has also drastically reduced the errors made by manual calculations.

Applications

Many potential applications have been suggested in literature, including enzyme immobilization, drug delivery systems, and nanotechnological self-assembly of materials. Though DNA is not the natural choice for building active structures for nanorobotic applications, due to its lack of structural and catalytic versatility, several papers have examined the possibility of molecular walkers on origami and switches for algorithmic computing. The following paragraphs list some of the reported applications conducted in the laboratories with clinical potential.

Researchers at the Harvard University Wyss Institute reported the self-assembling and self-destructing drug delivery vessels using the DNA origami in the lab tests. The DNA nanorobot they created is an open DNA tube with a hinge on one side which can be clasped shut. The drug filled DNA tube is held shut by a DNA aptamer, configured to identify and seek certain diseased related protein. Once the origami nanobots get to the infected cells, the aptamers break apart and release the drug. The first disease model the researchers used was leukemia and lymphoma.

Researchers in the National Center for Nanoscience and Technology in Beijing and Arizona State University reported a DNA origami delivery vehicle for Doxorubicin, a well-known anti-cancer drug. The drug was non-covalently attached to DNA origami nanostructures through intercalation and a high drug load was achieved. The DNA-Doxorubicin complex was taken up by human breast adenocarcinoma cancer cells (MCF-7) via cellular internalization with much higher efficiency than doxorubicin in free form. The enhancement of cell killing activity was observed not only in regular MCF-7, more importantly, also in doxorubicin-resistant cells. The scientists theorized that the doxorubicin-loaded DNA origami inhibits lysosomal acidification, resulting in cellular redistribution of the drug to action sites, thus increasing the cytotoxicity against the tumor cells.

In a study conducted by a group of scientists from iNANO center and CDNA Center at Aarhus university, researchers were able to construct a small multi-switchable 3D DNA Box Origami. The proposed nanoparticle was characterized by AFM, TEM and FRET. The constructed box was shown to have a unique reclosing mechanism, which enabled it to repeatedly open and close in response to a unique set of DNA or RNA keys. The authors proposed that this "DNA device can potentially be used for a broad range of applications such as controlling the function of single molecules, controlled drug delivery, and molecular computing."

Nanorobots made of DNA origami demonstrated computing capacities and completed pre-programmed task inside the living organism was reported by a team of bioengineers at Wyss Institute at Harvard University and Institute of Nanotechnology and Advanced Materials at Bar-Ilan University. As a proof of concept, the team injected various kinds of nanobots (the curled DNA encasing molecules with fluorescent markers) into live cockroaches. By tracking the markers inside the cockroaches, the team found the accuracy of delivery of the molecules (released by the uncurled DNA) in target cells, the interactions among the nanobots and the control are equivalent to a computer system. The complexity of the logic operations, the decisions and actions, increases with the increased number of nanobots. The team estimated that the computing power in the cockroach can be scaled up to that of an 8-bit computer.

DNA is folded into an octahedron and coated with a single bilayer of phospholipid, mimicking the envelope of a virus particle. The DNA nanoparticles, each at about the size of a virion, are able to remain in circulation for hours after injected into mice. It also elicits much lower immune response than the uncoated particles. It presents a potential use in drug delivery, reported by researchers in Wyss Institute at Harvard University.

Structural DNA Nanotechnology

Structural DNA nanotechnology, sometimes abbreviated as SDN, focuses on synthesizing and characterizing nucleic acid complexes and materials where the assembly has a static, equilibrium endpoint. The nucleic acid double helix has a robust, defined three-dimensional geometry that makes it possible to predict and design the structures of more complicated nucleic acid complexes. Many such structures have been created, including two- and three-dimensional structures, and periodic, aperiodic, and discrete structures.

Extended Lattices

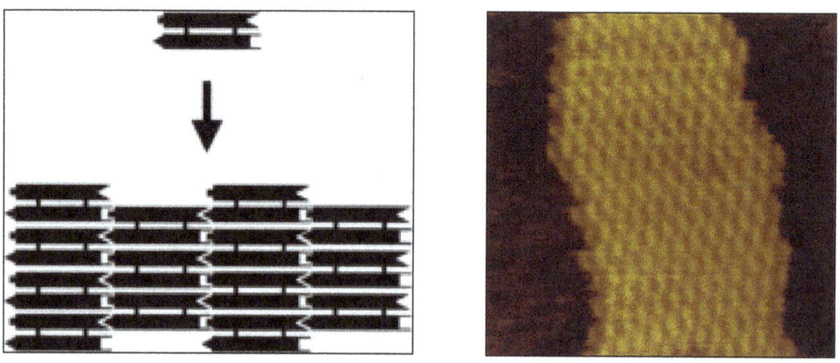

The assembly of a DX array. *Left*, schematic diagram.

Each bar represents a double-helical domain of DNA, with the shapes representing complementary sticky ends. The DX complex at top will combine with other DX complexes into the two-dimensional array shown at bottom. Right, an atomic force microscopy image of the assembled array. The individual DX tiles are clearly visible within the assembled structure. The field is 150 nm across.

Small nucleic acid complexes can be equipped with sticky ends and combined into larger two-dimensional periodic lattices containing a specific tessellated pattern of the individual molecular tiles. The earliest example of this used double-crossover (DX) complexes as the basic tiles, each containing four sticky ends designed with sequences that caused the DX units to combine into periodic two-dimensional flat sheets that are essentially rigid two-dimensional crystals of DNA. Two-dimensional arrays have been made from other motifs as well, including the Holliday junction rhombus lattice, and various DX-based arrays making use of a double-cohesion scheme. The top two images at right show examples of tile-based periodic lattices.

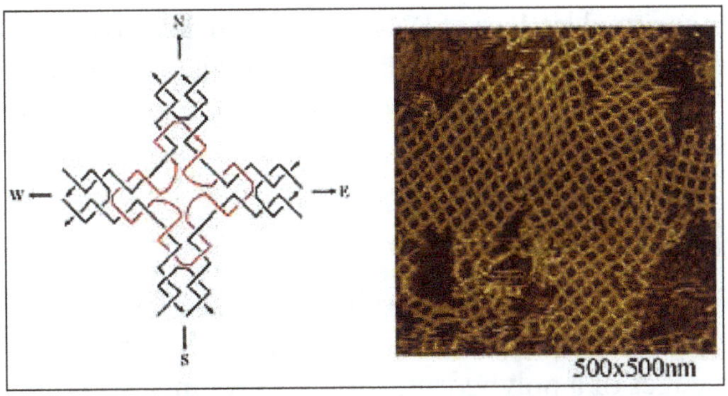

Left, a model of a DNA tile used to make another two-dimensional periodic lattice. Right, an atomic force micrograph of the assembled lattice.

Two-dimensional arrays can be made to exhibit aperiodic structures whose assembly implements a specific algorithm, exhibiting one form of DNA computing. The DX tiles can have their sticky end sequences chosen so that they act as Wang tiles, allowing them to perform computation. A DX array whose assembly encodes an XOR operation has been demonstrated; this allows the DNA array to implement a cellular automaton that generates a fractal known as the Sierpinski gasket. The third image at right shows this type of array. Another system has the function of a binary counter, displaying a representation of increasing binary numbers as it grows. These results show that computation can be incorporated into the assembly of DNA arrays.

DX arrays have been made to form hollow nanotubes 4–20 nm in diameter, essentially two-dimensional lattices which curve back upon themselves. These DNA nanotubes are somewhat similar in size and shape to carbon nanotubes, and while they lack the electrical conductance of carbon nanotubes, DNA nanotubes are more easily modified and connected to other structures. One of many schemes for constructing DNA nanotubes uses a lattice of curved DX tiles that curls around itself and closes into a tube. In an alternative method that allows the circumference to be specified in a simple, modular fashion using single-stranded tiles, the rigidity of the tube is an emergent property.

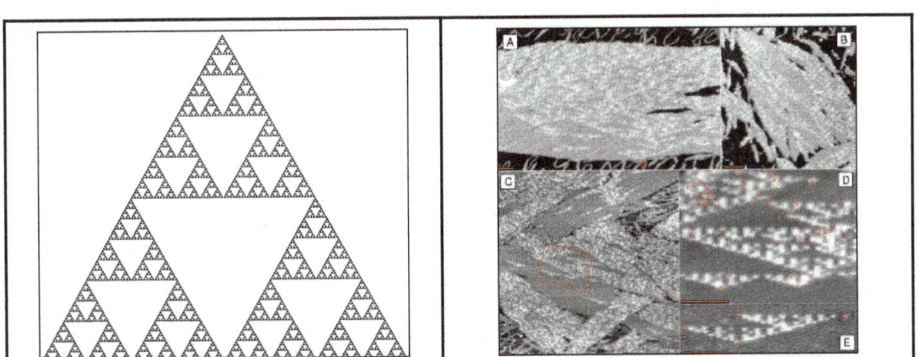

An example of an aperiodic two-dimensional lattice that assembles into a fractal pattern. Left, the Sierpinski gasket fractal. Right, DNA arrays that display a representation of the Sierpinski gasket on their surfaces.

Forming three-dimensional lattices of DNA was the earliest goal of DNA nanotechnology, but this proved to be one of the most difficult to realize. Success using a motif based on the concept of tensegrity, a balance between tension and compression forces, was finally reported in 2009.

Discrete Structures

Researchers have synthesized many three-dimensional DNA complexes that each have the connectivity of a polyhedron, such as a cube or octahedron, meaning that the DNA duplexes trace the edges of a polyhedron with a DNA junction at each vertex. The earliest demonstrations of DNA polyhedra were very work-intensive, requiring multiple ligations and solid-phase synthesis steps to create catenated polyhedra. Subsequent work yielded polyhedra whose synthesis was much easier. These include a DNA octahedron made from a long single strand designed to fold into the correct conformation, and a tetrahedron that can be produced from four DNA strands in one step.

Nanostructures of arbitrary, non-regular shapes are usually made using the DNA origami method. These structures consist of a long, natural virus strand as a "scaffold", which is made to fold into the desired shape by computationally designed short "staple" strands. This method has the advantages of being easy to design, as the base sequence is predetermined by the scaffold strand sequence, and not requiring high strand purity and accurate stoichiometry, as most other DNA nanotechnology methods do. DNA origami was first demonstrated for two-dimensional shapes, such as a smiley face, a coarse map of the Western Hemisphere, and the Mona Lisa painting. Solid three-dimensional structures can be made by using parallel DNA helices arranged in a honeycomb pattern, and structures with two-dimensional faces can be made to fold into a hollow overall three-dimensional shape, akin to a cardboard box. These can be programmed to open and reveal or release a molecular cargo in response to a stimulus, making them potentially useful as programmable molecular cages.

Templated Assembly

Nucleic acid structures can be made to incorporate molecules other than nucleic acids, sometimes called heteroelements, including proteins, metallic nanoparticles, quantum dots, and fullerenes. This allows the construction of materials and devices with a range of functionalities much greater than is possible with nucleic acids alone. The goal is to use the self-assembly of the nucleic acid structures to template the assembly of the nanoparticles hosted on them, controlling their position and in some cases orientation. Many of these schemes use a covalent attachment scheme, using oligonucleotides with amide or thiol functional groups as a chemical handle to bind the heteroelements. This covalent binding scheme has been used to arrange gold nanoparticles on a DX-based array, and to arrange streptavidin protein molecules into specific patterns on a DX array. A non-covalent hosting scheme using Dervan polyamides on a DX array was used to arrange streptavidin proteins in a specific pattern on a DX array. Carbon nanotubes have been hosted on

DNA arrays in a pattern allowing the assembly to act as a molecular electronic device, a carbon nanotube field-effect transistor. In addition, there are nucleic acid metallization methods, in which the nucleic acid is replaced by a metal which assumes the general shape of the original nucleic acid structure, and schemes for using nucleic acid nanostructures as lithography masks, transferring their pattern into a solid surface.

Dynamic DNA Nanotechnology

In addition to constructing complex static structures, DNA nanostructures can change state in response to interactions with each other and with their environment. These transitions can be used to create sensors, computers, and molecular machines.

Actuation and Response

Environmentally responsive structural motifs include guanineand cytosine-rich sequences that fold into stable quadruplexes in the presence of specific metal ions and low pH, ion- and temperature-sensitive base stacking, and sequences that can be switched from right- to left-handed. Single-stranded aptamers, produced by in vitro selection, can bind a wide variety of ligands. The corresponding structural changes can be used to create a sensor by, for example, modulating interactions between fluorophores or activating a catalyst. Light can trigger motion— planar azobenzene moieties intercalated between base pairs isomerize on ultraviolet illumination to destabilize the duplex; they can be switched back reversibly using visible light. Structures can also be changed by site-specific enzymatic ligation, cleavage, or degradation. Response times range from milliseconds for quadruplex formation to (typically) seconds or hours for enzymatic modification or optical actuation. Sensors can be interfaced directly to molecular computers and used to drive simple molecular machines.

Perhaps the most flexible means of inducing changes in nucleic acid nanostructures is the displacement of one oligonucleotide from a duplex by another. Strand displacement is frequently used for signaling and actuation; it can also be used to sense specific nucleic acid sequences for medical diagnosis or to interface natural cellular control mechanisms. The rate of strand invasion can be greatly increased by the use of "toeholds," single-stranded regions of the target complex to which the invader can bind to initiate the reaction forming additional base pairs that provide a thermodynamic driving force for invasion. Strand displacement reactions are relatively slow (typical systems take from seconds to hours), but a high degree of control can be achieved through sequence design. Strand-displacement reactions can join and release components, open and close cages, and change shapes; they form the basis of a system for molecular computation capable of calculating a square root.

Dynamic DNA systems often require repeated external stimulation to achieve cyclic motion. They can, however, be designed to run continuously, powered by chemical energy sources such as hybridization or enzymatic cleavage, or by light. Functions implemented by autonomous DNA machines include motion along self-assembled tracks and programmable chemical synthesis. The ability of nucleic acid nanostructures to operate autonomously and to sense, compute, and actuate suggests applications in medical diagnosis and locally controlled drug delivery; this is a strong theme in the development of the field.

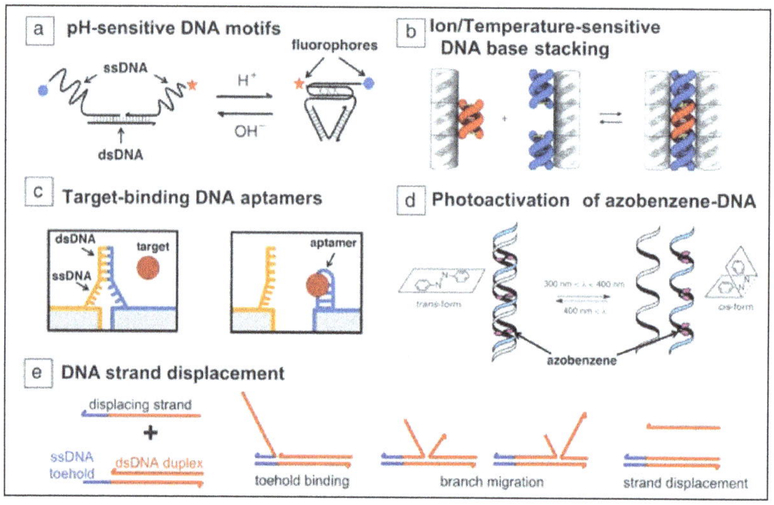

Examples of DNA nanostructure response elements.

Applications of DNA Profiling

The blueprint for our bodies is made up of genes, governing factors such as growth, development, and functioning. Almost every cell in the human body contains a copy of the blueprint, stored inside a special sac called the nucleus.

About 23,000 genes which are estimated are beaded along tightly bundled strands of a chemical substance called oligonucleotide acid (DNA). These strands are known as chromosomes.

Humans have 46 paired chromosomes (half inherited from each parent), with two sex chromosomes that decide gender and 44 chromosomes that dictate other factors.Some fragment of DNA is unique to each individual. DNA profiling is a way of estab-lishing identity and is used in a variety of ways, such as finding out whether twins are fraternal or identical. Samples are usually obtained from blood.

Uses of DNA Profiling

Some of the Uses of DNA Profiling Include:

- Paternity: To find out if the alleged father is actually the biological father of the child.

- The same genetic material is shared by identical twins, while fraternal (non-identical) twins develop from two eggs fertilized by two sperm and are no more alike than individual siblings born at different times. It can be difficult to tell at birth whether twins are identical or fraternal.
- Immigration: Some visa applications may depend on proof of relatedness.
- Justice in criminal cases: DNA testing can help solve crimes by comparing the DNA profiles of suspects to offender samples. Victorian law allows the collection of blood and saliva samples from convicted criminals and suspects. DNA profiles are then kept on a database.

Advantages of DNA profiling

A few advantages of DNA profiling include:

- Any human sample that contains cells with nuclei, such as saliva, semen, urine, and hair are applicable for DNA tests.
- These tests are extremely sensitive and can be conducted using samples that would be too small for other seismological tests.
- DNA do not degenerate even after contamination of chemicals in bacteria.
- The ability of DNA profiling to exclude a suspect means the police are able to confidently drop that line of inquiry and continue their investigation down other avenues.

Limitations of DNA Profiling

Contrary to public belief, DNA profiling isn't infallible. Critics point out various problems which are:

- Profiling technologies which are new can give incorrect results, due to errors such as cross-contamination of samples.
- Older DNA profiling technologies are more prone to errors, which could give false-negative or million), rather than absolute certainty. The more people tested, the lower the statistical probability.

Applications of DNA Sequencing

Forensics

DNA sequencing has been applied in forensics science to identify particular individual because every individual has unique sequence of his/her DNA. It is particularly used to identify the criminals by finding some proof from the crime scene in the form of hair, nail, skin or blood samples. DNA sequencing is also used to determine the paternity of the child. Similarly, it also identifies the endangered and protected species.

Medicine

In medical research, DNA sequencing can be used to detect the genes which are associated with some heredity or acquired diseases. Scientists use different techniques of genetic engineering like gene therapy to identify the defected genes and replace them with the healthy ones.

Agriculture

DNA sequencing has played vital role in the field of agriculture. The mapping and sequencing of the whole genome of microorganisms has allowed the agriculturists to make them useful for the crops and food plants. For example, specific genes of bacteria have been used in some food plants to increase their resistance against insects and pests and as a result the productivity and nutritional value of the plants also increases. These plants can also fulfill the need of food in poor countries. Similarly, it has been useful in the production of livestock with improved quality of meat and milk.

Applications of RNAi

The RNAi process holds the key to future technological applications. Genome sequencing projects generate a wealth of information. However, the ultimate goal of such projects is to accelerate the identification of the biological function of genes. The functions of genes can be analyzed with an appropriate assay, by examining the phenotype of organisms that contain mutations in the gene, or on the basis of knowledge gained from the study of related genes in other organisms. However, a significant fraction of genes identified by the sequencing projects are new and cannot be rapidly assigned functions by these conventional methods.

RNAi technology is proving to be useful to analyze quickly the functions of a number of genes in a wide variety of organisms. RNAi has been adapted with high-throughput screening formats in C. elegans, for which the recombination-based gene knockout technique has not been established. Chromosomes I and III of C. elegans have been screened by RNAi to identify the genes involved in cell division and embryonic development. Recently, a large-scale functional analysis of ≈19,427 predicted genes of C. elegans was carried out with RNA interference. This study identified mutant phenotypes for 1,722 genes. Similarly, in D. melanogaster, RNAi technology has been successfully applied to identify genes with essential roles in biochemical signaling cascades, embryonic development, and other basic cellular process. In plants, gene knockdown-related functional studies are being carried out efficiently when transgenes are present in the form of hairpin (or RNAi) constructs. Plant endotoxins could also be removed if the toxin biosynthesis genes are targeted with the RNAi constructs. Recently, the theobromine synthase of the coffee plant was knocked down with the hairpin construct of the transgene, leading to the production of decaffeinated coffee plants. Virus-induced gene silencing has also been proven to be a successful approach for plant genetics.

Given the fact that RNAi is easy to apply, whole-genome screens by RNAi may become a common method of choice in the near future. RNAi may facilitate drug screening and development by identifying genes that can confer drug resistance or genes whose mutant phenotypes are ameliorated by drug treatment, providing information about the modes of action of novel compounds. Although RNAi is unlikely to replace the existing knockout technology, it may have a tremendous impact for those organisms that are not amenable to the knockout strategy. It may also be a method of choice to study the simultaneous functions of a number of analogous genes in organisms in which redundancy exists with respect to a particular function, because many of these genes can be silenced simultaneously.

Given the gene-specific features of RNAi, it is conceivable that this method will play an important role in therapeutic applications. Since siRNAs direct cellular RNAi biology, these are potential therapeutic reagents because of their power to downregulate the expression pattern of mutant genes in diseased cells. However, central to this hypothesis is the assumption that the effect of exogenous siRNA applications will remain gene specific and show no nonspecific side effects relating to mismatched off-target hybridization, protein binding to nucleic acids, etc. Though it was demonstrated that mismatches of more than even one nucleotide within the 19- to 20-mer siRNAs effectively disrupted proper degradation of the target mRNA, the gene specificity of siRNAs needs to be confirmed on a genome-wide scale.

Recently, Chi et al. reported that the GFP siRNA-induced gene silencing of transient or stably expressed GFP mRNA was highly specific in the human embryonic kidney (HEK) 293 cell background. The specific silencing did not produce secondary changes in global gene expression, as detected by the DNA microarray experiment. They also failed to detect the presence of transitive RNAi in experimentally engineered human cell lines. In their own experiments, Semizarov et al. reached a similar conclusion while using siRNAs corresponding to akt1, rb1, and plk1 in the human non-small cell lung carcinoma cell line H1299. These experiments prove that siRNAs could be used as highly specific tools for targeted gene knockdown and can be used in high-throughput approaches and drug target validation. This exquisite sequence-specific effect of siRNAs has also been exploited in silencing the mutant allele of the diseased gene while not affecting the wild-type allele of the healthy version of the same gene.

siRNAs have been shown to inhibit infection by human immunodeficiency virus, poliovirus, and hepatitis C virus in cultured cell lines. Bitko and Barik successfully used siRNAs to silence genes expressed from respiratory syncytial virus, an RNA virus that causes severe respiratory disease in neonates and infants. siRNA treatment has also been shown to reduce the expression of the BCR-ABL oncoprotein in leukemia and lymphoma cell lines, leading to apoptosis in these cells. With respect to future medical applications, siRNA-based therapy seems to have a great potential to combat carcinomas, myeloma, and cancer caused by overexpression of an oncoprotein or generation of an oncoprotein by chromosomal translocation and point mutations.

Recently, the therapeutic potential of the siRNA technique has been demonstrated in vivo in mouse models. McCaffrey et al. and Song et al. demonstrated effective targeting of a sequence from hepatitis C virus and the fas gene by RNA interference in mouse liver. An epiallelic series of p53 hypomorphs created by RNAi have been shown to produce distinct tumor phenotypes in mice in vivo, suggesting that RNAi can stably suppress gene expression. Song et al. have shown that treatment with fas siRNA abrogated hepatocyte necrosis and inflammatory infiltration and protected mice from liver fibrosis and fulminant hepatitis. Rubinson et al. showed highly specific, stable, and functional silencing of gene expression in transgenic mice with the lentivirus system for the delivery of siRNAs.

Although the delivery of siRNAs to a proper site remains problematic for gene therapy, chemical modifications of siRNAs such as changing the lipophilicity of the molecules or the methods previously developed for the application of antisense oligonucleotides or nuclease-resistant ribozymes might help the entry and stability of siRNAs within the transfected cells or tissues. The absence of specific micro-RNAs has been demonstrated in carcinoma cells, implying that cancer development could be arrested by introduction of the missing micro-RNAs. The micro-RNAs could be supplied in the form of siRNAs, since the function of micro-RNAs can be mimicked by the exogenous siRNA. However, independent of its biomedical applications, RNAi appears to be a forthcoming method for functional genomics.

Medical Applications of RNA Interference

Medical sciences have undergone several changes of paradigms. In this quest to develop more potent drugs, less invasive surgical techniques and more sophisticated diagnostic approaches (among other goals) medicine has contributed decisively to our well being and longevity. Some of these major accomplishments, which altered the course of humanity are: the development of antibiotics, organ transplantation and radiography. Medicine has recently reached a new hallmark: RNA interference (RNAi), a Nobel prize winning technology which promises to promote medical care at the molecular level, by regulating genes in our favor.

RNAi was first successfully used in C. elegans, a laboratory model species, by microinjecting dsRNAs in the worm's gonad. However, the works that laid the foundations for the development of the technique were initially done in plants and fungi. RNAi was accidentally discovered, when white transgenic petunias were obtained instead of deep purple ones; a similar paradoxal result was obtained when white transgenic colonies of Neurospora sp were observed in place of orange ones. These findings revealed the existence of an unknown potent mechanism of gene regulation, which could be tamed for diverse uses, from basic science, agriculture, animal health and medicine.

From these seminal observations, during a period of nearly 10 years, researchers aimed at characterizing, at the genetic and biochemical levels, the process of RNAi. By the year 2001, scientists were ready to perform the initial analyses in human cells. Since then, RNAi has been used as a tool in molecular biology, allowing potent and specific gene silencing. This is achieved by the introduction of double-stranded RNA molecules (dsRNAs), which ultimately lead to the degradation of messenger RNAs (mRNAs), i.e., gene silencing. The molecular mechanism behind RNAi requires the activation of an endogenous pathway which is present in almost all eukaryotic cells. This pathway is responsible for detecting long dsRNAs and, subsequently, promoting their cleavage by an enzyme called dicer, into small duplexes of approximately 21-25 bp called siRNAs (small interfering RNAs). Alternatively, siRNAs can be directly introduced into the cytoplasm of eukaryotic cells. This second approach is required in mammalian cells, since dsRNAs longer than 30 bp induce apoptosis. siRNAs are subsequently transferred to a complex called RISC (RNA-induced silencing complex), where one strand of the duplex is destroyed and the other remains intact (designated guide strand). RISC then searches the cytoplasm for mRNA molecules with sequence complementarity to the guide strand. Once a mRNA with full complementarity is found, slicer (an endoribonuclease which is part of RISC) performs the cleavage of target mRNA.

References

- Genetic-engineering-dna-technology-applications, genetics-evolution, biology, math-science: infoplease.com, Retrieved 28 May, 2020
- Garde, Damian (May 15, 2012). "DNA origami could allow for 'autonomous' delivery". fiercedrugdelivery.com. Retrieved May 25, 2012
- Recombinant-DNA-Technology: mybiosource.com, Retrieved 13 March, 2020
- Background: Pelesko, John A. (2007). Self-assembly: the science of things that put themselves together. New York: Chapman & Hall/CRC. pp. 5, 7. ISBN 978-1-58488-687-7
- Applications-of-Recombinant-DNA-Technology-in-Medicine, Genetics-Article-350: biotecharticles.com, Retrieved 30 August, 2020
- DNA walkers: Shin, Jong-Shik; Pierce, Niles A. (8 September 2004). "A synthetic DNA walker for molecular transport" (PDF). Journal of the American Chemical Society. 126(35): 10834–10835. doi:10.1021/ja047543j. PMID 15339155
- Practical-aspects-of-structural-and-dynamic-DNA-nanotechnology-321675577: researchgate.net, Retrieved 15 June, 2020

Permissions

All chapters in this book are published with permission under the Creative Commons Attribution Share Alike License or equivalent. Every chapter published in this book has been scrutinized by our experts. Their significance has been extensively debated. The topics covered herein carry significant information for a comprehensive understanding. They may even be implemented as practical applications or may be referred to as a beginning point for further studies.

We would like to thank the editorial team for lending their expertise to make the book truly unique. They have played a crucial role in the development of this book. Without their invaluable contributions this book wouldn't have been possible. They have made vital efforts to compile up to date information on the varied aspects of this subject to make this book a valuable addition to the collection of many professionals and students.

This book was conceptualized with the vision of imparting up-to-date and integrated information in this field. To ensure the same, a matchless editorial board was set up. Every individual on the board went through rigorous rounds of assessment to prove their worth. After which they invested a large part of their time researching and compiling the most relevant data for our readers.

The editorial board has been involved in producing this book since its inception. They have spent rigorous hours researching and exploring the diverse topics which have resulted in the successful publishing of this book. They have passed on their knowledge of decades through this book. To expedite this challenging task, the publisher supported the team at every step. A small team of assistant editors was also appointed to further simplify the editing procedure and attain best results for the readers.

Apart from the editorial board, the designing team has also invested a significant amount of their time in understanding the subject and creating the most relevant covers. They scrutinized every image to scout for the most suitable representation of the subject and create an appropriate cover for the book.

The publishing team has been an ardent support to the editorial, designing and production team. Their endless efforts to recruit the best for this project, has resulted in the accomplishment of this book. They are a veteran in the field of academics and their pool of knowledge is as vast as their experience in printing. Their expertise and guidance has proved useful at every step. Their uncompromising quality standards have made this book an exceptional effort. Their encouragement from time to time has been an inspiration for everyone.

The publisher and the editorial board hope that this book will prove to be a valuable piece of knowledge for students, practitioners and scholars across the globe.

Index

www.ingramcontent.com/pod-product-compliance
Lightning Source LLC
LaVergne TN
LVHW080634210926
848485LV00003B/261
9781639871681